WATER FOR THE

FUTURE

THE WEST BANK AND GAZA STRIP, ISRAEL, AND JORDAN

D1104665

Committee on Sustainable Water Supplies for the Middle East

Israel Academy of Sciences and Humanities
Palestine Academy for Science and Technology
Royal Scientific Society, Jordan
U.S. National Academy of Sciences

NATIONAL ACADEMY PRESS
Washington, D.C. 1999

NOTICE: The project that is the subject of this report was approved by the Governing Board of the National Research Council and by competent authorities of the Israel Academy of Sciences and Humanities, the Royal Scientific Society of Jordan, and the Palestine Academy for Science and Technology. The members of the committee responsible for the report were chosen for their special competences and with regard for appropriate balance.

This study was supported by the Casey Program Fund and the Arthur Day Program Fund of the U.S. National Research Council Fund. Any opinions, findings, conclusions, or recommendations expressed in this publication are those of the author(s) and do not necessarily reflect the view of the organization that provided support for this project.

Library of Congress Catalog Card Number 97-80489
International Standard Book Number 0-309-06421-X

Additional copies of this report are available from:

National Academy Press
2101 Constitution Avenue, NW
Box 285
Washington, DC 20055
800-264-6242
202-334-3313 (in the Washington metropolitan area)
http://www.nap.edu

Cover art: Watercolor by Marc Castelli, Chestertown, Md., after photograph by David Policansky.

Printed in the United States of America.

ISRAEL ACADEMY OF SCIENCES AND HUMANITIES

The Israel Academy consists of seventy of Israel's most distinguished scientists and scholars. It is managed by a council consisting of a president, vice-president, chairmen for the sciences and the humanities, and an executive director. The president is appointed directly by the president of the state, upon the recommendation of the Academy council. Professor Jacob Ziv is the current president. The governing regulations of the Academy are self-determined, in consultation with the minster of education and culture. The Academy is divided into two major sections: natural sciences and humanities. Both share equal responsibility in conducting the Academy's affairs.

The Israel Academy represents the Israeli scientific community in national and international fora. It also initiates and sponsors selected scholarly projects, conferences and publications, often in conjunction with academic bodies abroad. The Academy also founded and provides administrative support to the Israel Science Foundation (ISF), and administers several other privately sponsored national grants and scholarship programs.

PALESTINE ACADEMY FOR SCIENCE AND TECHNOLOGY

The Palestine Academy for Science and Technology is an autonomous, government-supervised, non-profit organization that is mandated to represent the Palestinian national instrument for making strategic investment in Palestine's capability and potential in science and technology, in that science and technology should impact the Palestinian national economy and improve the quality of life for all Palestinians. As such, the Academy represents the Palestinian commitment to respond to the science and technology and research and development needs of the Palestinian private and public sectors. In this role, the Academy represents the focal point to a wide scientific and technological Palestinian network that is geared towards capacity building of the Palestinian R&D; promotion and application of science and technology as a tool for socio-economic development; and advising the government on science and technology issues.

ROYAL SCIENTIFIC SOCIETY, JORDAN

The Royal Scientific Society (RSS) was established in 1970 in accordance with a Royal Decree. Since its inception, His Royal Highness Crown

Prince El-Hassan has sponsored its march and was chairman of its Board of Trustees.

In April 1987 the Higher Council for Science and Technology (HCST) was established under the chairmanship of His Royal Highness Crown Prince El-Hassan. HCST replaced the RSS Board of Trustees and RSS became one of its affiliated institutions.

RSS is a nonprofit institution enjoying financial and administrative independence. It aims at conducting scientific and technological research and development work related to the development process in Jordan with special attention to industrial research and services. It also aims at disseminating awareness in the scientific and technological fields and at providing specialized technical consultations and services to the public and private sectors. It seeks to develop scientific and technological cooperation with similar institutions within the Arab world and internationally.

U.S. NATIONAL ACADEMY OF SCIENCES

The National Academy of Sciences is a private, nonprofit, self-perpetuating society of distinguished scholars engaged in scientific and engineering research, dedicated to the furtherance of science and technology and to their use for the general welfare. Upon the authority of the charter granted to it by the Congress in 1863, the Academy has a mandate that requires it to advise the federal government on scientific and technical matters. Dr. Bruce M. Alberts is president of the National Academy of Sciences.

The National Academy of Engineering was established in 1964, under the charter of the National Academy of Sciences, as a parallel organization of outstanding engineers. It is autonomous in its administration and in the selection of its members, sharing with the National Academy of Sciences the responsibility for advising the federal government. The National Academy of Engineering also sponsors engineering programs aimed at meeting national needs, encourages education and research, and recognizes the superior achievements of engineers. Dr. William A. Wulf is president of the National Academy of Engineering.

The Institute of Medicine was established in 1970 by the National Academy of Sciences to secure the services of eminent members of appropriate professions in the examination of policy matters pertaining to the health of the public. The Institute acts under the responsibility given to the National Academy of Sciences by its congressional charter to be an adviser to the federal government and, upon its own initiative, to identify issues of medical care, research, and education. Dr. Kenneth I. Shine is president of the Institute of Medicine.

The National Research Council was organized by the National Academy of Sciences in 1916 to associate the broad community of science and technology with the Academy's purposes of furthering knowledge and advising the federal government. Functioning in accordance with general policies determined by the Academy, the Council has become the principal operating agency of both the National Academy of Sciences and the National Academy of Engineering in providing services to the government, the public, and the scientific and engineering communities. The Council is administered jointly by both Academies and the Institute of Medicine. Dr. Bruce M. Alberts and Dr. William A. Wulf are chairman and vice chairman, respectively, of the National Research Council.

Acknowledgments

Many individuals assisted the committee in its task by participating in committee meetings, helping to plan field trips, and providing background information. The committee is especially thankful to the participants from the science academies and councils who met September 4-5, 1995 in Amman, Jordan, and agreed upon a scope of work. They are as follows:

Mr. Sami Abbasi
Ministry of Water and Irrigation, Jordan

Dr. Ali Al Ghazawi
Royal Scientific Society, Jordan

Dr. Bruce M. Alberts, President
U.S. National Academy of Sciences

Dr. Lindsay H. Allen
University of California, Davis

Dr. Sa'id Alloush, President
Royal Scientific Society, Jordan

Dr. Fathi Arafat, President
Palestine Higher Health Council

Dr. Khaled Elshuraydeh
Deputy Secretary General
Higher Council for Science and Technology, Jordan

Dr. David L. Freyberg, Stanford University
Chair, Water Science and Technology Board (July 1994–June 1997)
U.S. National Research Council

Mr. Ahmed Hatamleh
Ministry of Water and Irrigation, Jordan

Dr. Naim A. Ismail, Executive Director
Environmental Health Central Unit
Palestinian Higher Health Council

Professor Ibrahim Khatib
Jordan University for Science and Technology

Dr. Hani Mulki, President (May 1989–June 1997)
Royal Scientific Society of Jordan

Dr. F. Sherwood Rowland, Foreign Secretary
U.S. National Academy of Sciences

Dr. Kenneth I. Shine, President
Institute of Medicine
U.S. National Academy of Sciences

Professor Menahem Yaari, Vice President
Israel Academy of Sciences and Humanities

Dr. Meir Zadok, Director
Israel Academy of Sciences and Humanities

The committee would also like to express appreciation to others who attended committee meetings and provided background information, those who assisted in making arrangements for the field trips, and the hosts for the committee meetings. They are:

February 14-16, 1996, Meeting
Washington, D.C.

Dr. James Crook, Black & Veatch Consulting Engineers, Cambridge, Massachusetts, USA

Dr. Jeffrey Goodsen, Agency for International Development, Washington, D.C., USA

Dr. Daniel Okun, University of North Carolina at Chapel Hill, USA

Dr. Alexander McPhail, World Bank, Washington, D.C., USA

Dr. Roy Popkin, Consultant, Silver Spring, Maryland, USA

Dr. Aaron Wolf, University of Alabama, Tuscaloosa, USA

June 17-19, 1996, Meeting
Amman, Jordan

Dr. Munther Haddadin, Royal Scientific Society, Amman, Jordan

Dr. Uri Shamir, Water Research Institute, Technion, Israel Institute of Technology, Haifa

Participants for Field Trip

Mr. Mohammed Abu Taha, ZAI Water Works

Mr. Faza Saleh, King Abdullah Canal

Hosts

Dr. Hani Mulki, President, Royal Scientific Society, Amman, Jordan (1989-1997)

Dr. Sa'id Alloush, Vice President, Royal Scientific Society, Amman, Jordan

Ms. Samar Al-Rabadi, Royal Scientific Society, Amman, Jordan

September 18-20, 1996, Meeting
Haifa, Israel

Participants for Field Trip

Dr. Aric Belkind, Regional Water Council

Dr. Jonathan Ben Zur, Mekoroth Water Company, Ltd.

Dr. Giora Shacham, Tznobar Company

Hosts

Dr. Yoram Avnimelech, Technion, Israel Institute of Technology, Haifa
Ms. Sandra Hesseg, Water Research Institute, Technion, Israel Institute
     of Technology
Professor Israela Ravina, Technion, Israel Institute of Technology, Haifa
Dr. Uri Shamir, Water Research Institute, Technion, Israel Institute of
     Technology, Haifa
Dr. Zehev Tador, President, Technion, Israel Institute of Technology,
     Haifa
Dr. Meir Zadok, Israel Academy of Sciences and Humanities, Jerusalem,
     Israel

REVIEWERS

This report has been reviewed by individuals chosen for their diverse perspectives and technical expertise, in accordance with procedures approved by the NRC's Report Review Committee. The purpose of this independent review is to provide candid and critical comments that will assist the authors and the NRC in making the published report as sound as possible and to ensure that the report meets institutional standards for objectivity, evidence, and responsiveness to the study charge. The content of the review comments and draft manuscripts remain confidential to protect the integrity of the deliberative process. We wish to thank the following individuals for their participation in the review of this report:

Khairy Al Jamal, Palestinian Water Authority
Kenneth J. Arrow, Stanford University, Stanford, California
Nathan Buras, University of Arizona, Tucson
Ingrid C. Burke, Colorado State University, Ft. Collins
Paul L. Busch, Malcolm Pirnie, Inc., White Plains, New York
John Cairns, Virginia Polytechnic Institute and State
     University (Emeritus)
Hazim El-Naser, Ministry of Water and Irrigation, Amman, Jordan
Wilford R. Gardner, University of California, Berkeley (Retired)
Philip E. LaMoreaux, LaMoreaux & Associates, Inc., Tuscaloosa,
     Alabama
Thomas C. Schelling, University of Maryland, College Park
Yossi Segal, Israel Academy of Sciences and Humanities
A. Dan Tarlock, IIT Chicago Kent College of Law, Chicago, Illinois

While the individuals listed above provided many constructive comments and suggestions, responsibility for the final content of this report rests solely with the authoring committee and the NRC.

Preface

Representatives of the principal science councils of Israel, Jordan, the Palestinian Authority, and the United States first met in Washington, D.C., in 1994, to consider ways in which they might collaborate for the mutual benefit of their communities. After canvassing a range of problems of vital and common interest, they concluded that the most critical of these problems was ensuring sustainable water supplies in the Middle East. They decided to develop a joint study using the approach of the National Research Council (e.g., use of volunteer, multidisciplinary experts to write consensus reports on science and technology issues). They then appointed a committee with members from Israel, Jordan, the West Bank and Gaza Strip, the United States, and Canada. This report is the result.

The Committee on Sustainable Water Supplies for the Middle East was formed by the U.S. National Research Council of the U.S. National Academy of Sciences, National Academy of Engineering, and the Institute of Medicine; the Royal Scientific Society, Jordan; the Palestine Health Council; and the Israel Academy of Sciences and Humanities and has been supported by funds from the U.S. National Research Council. The plans for this study were developed during a meeting of the scientific Academies and Councils of the Middle East Region in Amman, Jordan in 1995.

There was little precedent for the committee's specific task. Specifying the criteria for sustainable development of an area's water resources, including the maintenance of natural support systems, was relatively new.

Additionally, identifying the scientific and research components of such an appraisal without linking them to specific development plans and allocations was also new. Moreover, this study was one of few occasions on which the U.S. National Research Council had ventured such a cooperative undertaking with scientists in other countries. And for Israel, Jordan, and the Palestinian Authority organizations, the study was unprecedented. While representatives of natural science academies had consulted briefly on such issues as population policy, this was a new undertaking. The standard review process of the U.S. National Research Council was modified to include the presidents of all four scientific organizations and other scientists from all four countries.

Notwithstanding a good deal of discussion in international scientific circles, there have been very few attempts to apply definitions and measurements of indices of sustainable development of water resources in a unified fashion to one area, as is documented in this report. Reports resulting from many of the previous management studies are listed in the bibliography. There has also been a notable lack of integration of social and economic considerations with considerations of ecosystem health and services, especially as they are related to biodiversity.

The committee has sought to integrate these considerations in its approach to this study. The study area is defined as all parts of Israel, Jordan, and the West Bank and Gaza Strip. Although other countries east of the Mediterranean, often encompassed by the term "Middle East," have somewhat similar landscapes and hydrologic features, it was necessary to confine the study to this area as lying wholly within the committee's concern.

The committee sought to canvass the full range of alternatives—physical, biological, and social that might be considered in sustaining the water supplies of the study area. Each option was then assessed with regard to its likely effects on the sustainability of the systems involved. Options that are currently practicable were distinguished from those that might be achieved through further research. Additionally, the committee has applied the basic principle of seeking fairness to both present and future generations by considering intergenerational equity in all of its appraisal.

While the committee was aware of many specific proposals for managing water supplies in the study area and neighboring areas, it explicitly refrained from making recommendations about specific development programs and political policies. The committee did attempt to appraise the state of scientific and technologic knowledge of water supplies that would be basic for any sound management program.

The conduct of the study included meetings in Israel, Jordan, and the United States, and field inspections of representative areas such as the Yarmuk and Jordan valleys and the Hula Valley. It also drew on the written and oral contributions from many people familiar with water resources and problems throughout the study area. Careful attention was paid to the substantial bodies of research findings from universities and research groups in the cooperating countries and other areas, including those named in the acknowledgments and bibliography.

While during the period of the study the international political scene in the study area was marked by tensions and contending charges, this situation did not color or interfere with the participation of committee members or scientific agencies from which they drew information and expert opinion. This final report was unanimously approved by all committee members and reflects a friendly process of frank discussion and mutual learning that it is hoped will be continued by scientists and scholars. This cooperative spirit was seen even in the unprecedented process of peer review by representatives of four political communities.

The frank, constructive tone of the committee deliberations would not have been possible without the sympathetic support of the members of the cooperating institutions and individuals named in the acknowledgments, and of a committee staff deeply respected for its scientific rigor and administrative competence. These staff were Sheila David and David Policansky, who had the administrative assistance of Jeanne Aquilino and the collaboration of WSTB director Stephen Parker. Individual committee members participated in preparing this final document in a variety of ways. All shared the sense that this was a unique opportunity, a chance to demonstrate the ability of concerned scientists and engineers to jointly help lay the groundwork for peaceful solutions to issues of critical social and environmental import in the foreseeable future.

Gilbert F. White, *Chair*, Committee on
Sustainable Water Supplies for the Middle East

Contents

Executive Summary

Sustaining the freshwater resources of the Middle East presents a great challenge and opportunity for the region's scientific and technological communities. They must play a fundamental role in ensuring that this essential but scarce resource is available for present and future generations. The challenge has been heightened, as problems with freshwater quality and availability have multiplied and changed in response to growing population and economic activity over the past several decades.

Adequate water supplies to meet basic human needs are essential to maintain and enhance the welfare of all the inhabitants of the region. For the present generation, water-related concerns primarily focus on the distribution of the resource within society and the preservation and protection of water quality. For future generations, additional concerns will be to ensure adequate water supplies and preserve the quality of the environment, in addition to achieving greater equity in the distribution of water throughout the area.

In focusing on the contributions of science and technology to the sustainable use of the study area's water resources, this study was guided by five fundamental working criteria:

1. The view taken should be regional.

2. The demands and needs of both present and future generations must be taken into account.

3. All options should be considered for balancing water supplies and demand.

4. The maintenance of ecosystem services should be viewed as essential for achieving sustainability of water resources.

5. The close relationships of water quality and quantity should be clearly recognized.

There is an important concept embodied in the terms sustainability and intergenerational equity—the idea that the present generation's children and grandchildren should have at least as much ability to use a resource as does the present generation. Intergenerational equity includes the sustainable use of water resources.

THE STUDY AREA AND WATER USE

The committee's deliberations were limited to the area of the West Bank and Gaza Strip, Israel, and Jordan, referred to as "the study area" in this report. The study area has a hot, dry climate, and consists of a dry coast and strip of dry upland forest that grades into semidesert and desert. Most of the study area receives less than 250 millimeters (mm) of rainfall per year, about the same as or less than that received by Phoenix, Arizona, in the United States. The study area's highest rainfall amounts—those of more than 1,000 mm—fall in a small area of highlands in the northwestern part of the study area. By comparison, most of the United States east of the Mississippi River receives more than 800 mm of rainfall per year; much of the eastern United States and much of the Pacific Northwest west of the Cascades receive 1,000 mm or more. The landscape and hydrologic features of the study area are much like those of neighboring areas, which are sometimes included in definitions of the "Middle East," stretching as far south as Yemen, as far east as Pakistan, as far north as Turkey, and as far west as Morocco.

The study area has approximately 12 million inhabitants, with varying proportions in urban centers and holding a variety of occupations. In 1994, the study area's total average annual water use was estimated to be 3,183 million cubic meters (million m³), ranging from almost 2,000 million m³ in Israel to approximately 235 million m³ in the West Bank and Gaza Strip. The average annual per capita use in the area, while highly variable, was approximately 260 cubic meters in 1994, and has been increasing. For the study area as a whole, agricultural irrigation accounts for more than half the water use, from an estimated 57 percent in Israel to 72 percent in Jordan, without considering wastewater that may be reused for irrigation. The several problems of water and the environment are similar to those in some neighboring areas and in some distant regions, such as arid sections of the United States and Australia.

Long experience in predicting water use and associated economic activity, population growth, and other variables of importance to water and economic planning shows that precise predictions are often incorrect. Many factors that influence water use have their origins outside the region, as described in Chapter 3, and even factors within the region can be unpredictable. However, although predictions, projections, and scenario building rarely provide an adequate basis for planning by themselves, they can be useful in identifying and analyzing different options.

The study area's inhabitants will almost assuredly live under conditions of significant water stress in the near future. Barring completely unforeseen events, the population is likely to grow, and very rapid growth is possible. The study area will probably continue to develop economically as well, and such growth could be substantial in Jordan and the West Bank and Gaza Strip. Because of the disparity between the economic progress of Israel compared to Jordan and the West Bank and Gaza Strip, some of the technical water-conservation and supply-augmentation options may be phased in over a long period of time. Chapter 2 provides a detailed discussion of the study area and its patterns of water use.

WATER AND THE ENVIRONMENT

The importance of ecosystem services to the sustainability of water supplies is often overlooked in the context of the region's water supplies. *Ecosystem services* refers to any functional attribute of natural systems that is beneficial to human society, nature, and wildlife. A biologically impoverished natural system produces services of poorer quality and reduced quantity.

The ecosystems of the study area, as elsewhere, provide services that are instrumental in achieving the sustainability of human water supplies. Vegetation helps to control runoff, and many plants, especially in wetlands, help to filter water and reduce the adverse effects of floods. Plants help to reduce erosion by reducing the rate of surface flows after heavy rains, thus reducing sediment input to water supplies as well. Surface water also provides important services. Streams help to assimilate wastewater, lakes provide storage for clean water, and surface waters provide habitat for many plants and animals important to humans and to ecosystem functioning.

At the same time, all the region's ecosystems, terrestrial as well as aquatic, require water for their own sustainability to continue to provide the many ecosystem goods and services that people rely on. The sustainability of water supplies requires that natural ecosystems be regarded as a critical legitimate user of the study area's water resources. Natural ecosystems are essential for maintaining adequate supplies of high-quality water.

Biodiversity is also important (see Chapter 4). Many peoples, including those in the Middle East, are committed to protecting biodiversity, as reflected in laws and international agreements. In future land-use planning, the benefits of water-related development should therefore be evaluated against the lost biodiversity and the cost of reduced ecosystem services. Applying this approach to the Jordan River basin as a whole would mean examining the effects of proposed measures on the biodiversity of wetlands, lakes, the lower river, and the Dead Sea coasts. Such an examination, lacking to date, should be an integral part of evaluating any proposed option that would affect water quantity or quality.

Chapter 4 discusses these evironmental issues in depth. In short, without the services provided by natural ecosystems, it will be extremely difficult and expensive—perhaps impossible—to sustain high-quality water supplies for the people in the study area. Thus, environmental considerations are not an adjunct to planning for sustainable water supplies, but a major and essential component of such planning.

HYDROLOGIC RELATIONSHIPS AND
WATER RESOURCES PLANNING

Regardless of national boundaries, the waters of the study area are shared inasmuch as the region is hydrologically connected. Changes in the quantities and qualities of water available in one area will have impact on the quantities and qualities available in others. A good way to ensure that these consequential relationships are directly considered in water resources planning is to take a regional hydrologic viewpoint. For example, the failure to view water resources planning regionally could lead to indiscriminate ground-water development of the Mountain Aquifer that underlies both Israel and the West Bank. Systematic determination of well locations would maximize the yield of this aquifer. A comprehensive hydrologic database to inform and support regional water resources planning is clearly needed.

It is recommended that responsible national and international agencies take a regional approach to water resources planning in the following fundamental ways:

1. Acquire data on water availability and water use by employing consistent methods, techniques, and protocols.
2. Monitor both quantity and quality conditions of the area's water resources using these consistent techniques and units of measurement.
3. Encourage open exchange of scientific research relevant to these water resources and the conduct of scientific research on a regional and collaborative basis.

Any regional approach should be cognizant of human equity and established legal water rights of shared resources.

SELECTED OPTIONS FOR THE FUTURE

Achieving intergenerational fairness implies the need for a variety of management measures, some discussed in this report. These measures include monitoring the quality of water resources; scientific and technological research and development to make more efficient use of available resources without contaminating or degrading the resource; intergenerational assessments of the effects of particular water projects and uses; effective maintenance of capital investments, such as dams, municipal sewage treatment plants and water delivery systems; protection of watersheds and aquifer recharge areas by appropriate land use planning; and systems for sharing the resources equitably among communities.

This report assesses specific management options to shape the study area's future water resources and use, keeping in mind the criteria noted previously (see page 1). Some of these options have received close attention; others have been examined only in part. (The bibliography found at the end of the report indicates the wide range of evidence the committee consulted.)

Most of the options examined here relate to improving the efficiency of water use—that is, they involve conservation and better use of proven technologies. Although new technologies hold some promise for increasing water supplies, none currently appears to be cost-effective and ready for large-scale application. The committee did not consider options that involved water sources outside the study area because this examination was outside its charge outlined in Chapter 1.

The committee identified several basic questions to consider when choosing among various water resource planning options:

1. How effective will the option be in enhancing available water supplies? Options that produce large increments in available water supply will be more desirable than those that have modest effects.

2. Are the options technically feasible? In evaluating options, care should be taken to assess technical feasibility.

3. What is the environmental impact of the option? Will the option reduce or increase the quantities or qualities of water supply for other uses? Does the option have any other adverse environmental impacts? How will it affect aquatic and terrestrial habitats? Will the option lead to losses of biodiversity or of species that may be particularly valuable?

4. Is the option economically feasible? What factors affect its economic feasibility? Has the option proven economically feasible else-

where? It is important in answering these questions that all costs and benefits be reported together, with appropriate information about who bears the costs and who receives the benefits.

5. What are the implications for present and future generations? The quality of the environment must be maintained for future generations in a condition no worse than that of the current generation. Will the current generation's access to resources be conserved for future generations?

In examining the options the committee explicitly considered these questions. However, in assessing options for a particular case, it will become more critical to examine and compare the full range of options suitable for that case. All too often, a proposed solution is examined according to only one criterion, such as monetary cost, and is not compared with the other possible actions. Because sustaining high-quality water supplies in the area will be extremely difficult and expensive without the goods and services provided by natural ecosystems, environmental considerations are essential in planning for sustainable water supplies (see Chapter 4). Thus, attempting to meet future regional demands by simply increasing withdrawals of ground and surface water will result in unsustainable development characterized by widespread environmental degradation and depletion of freshwater resources.

Conservation

Given the rate of population growth, water quality and quantity will not be sustainable unless suitable conservation methods are used in all three major sectors of water use—urban, agricultural, and industrial. Some middle ground must be reached in which quality of life and economic development are brought into balance within the practical constraints imposed by the available water. Measures to reduce the demand for water are generally well established, but often require societal or economic incentives to implement. By reducing the demand for water, conservation measures can have a positive effect on water quality and the environment.

Examples of voluntary, domestic water conservation measures include adopting water-saving plumbing fixtures (toilets, showerheads, and washing machines); limiting outdoor uses of water, as by watering lawns and gardens only during the evening and early morning; adopting water-saving practices in commerce; repairing household leaks; and discouraging use of garbage disposal units. Chapter 5 compares the water savings of nonconventional over conventional appliances. Involuntary domestic water conservation measures can also be used, such as repairing leaking distribution and sewer systems; expanding central sewage systems; me-

tering all water connections; and rationing, restricting, and recycling water use.

To the extent that the population grows or relocates in clusters, new
water systems can be designed to reduce use and treatment costs, as by
incorporating dual water systems to use nonpotable water for toilet flushing, garden irrigation, and similar applications. Dual systems reduce
treatment costs and allow for recycling.

Agriculture

The agricultural sector is the largest user of water in the study area.
Conservation measures have already helped to reduce the area's agricultural water use. The reduction of Israeli water use by more than 200
million m^3/yr between 1985 and 1993 was accomplished almost entirely
in the agricultural sector through improved irrigation methods and wa-
ter-delivery restrictions. Through rationing, research, and possibly
through economic policies, agricultural water use may become even more
efficient. However, as regional nonagricultural water demand increases
and the cost of obtaining additional water supplies grows more expensive, the role of agriculture in the economy of the study area may need to
be reevaluated (e.g., shifting from more to less water-intensive crops), so
that water is used as efficiently as possible.

Harvesting local runoff and floodwaters can increase water supplies
for dryland agriculture, and evaporative water loss can be reduced by
cropping intensively within closed environments. Computer-controlled
drip "fertigation" (application of fertilizer in the irrigation water) economizes on water and fertilizer use and prevents soil salinization and
ground-water pollution if drainage water recycling is used. Brackish
water, often abundant in the study area's dryland aquifers, can also be
used for irrigating salinity-resistant crops, increasing the sugar contents
of fruits such as tomatoes and melons and hence their market price. Brackish water is also useful for intensive aquaculture in deserts, but it may
also cause problems by increasing soil salinization.

Finally, the use of treated local or transported wastewater for subsurface drip irrigation of orchards and forage can dramatically increase the
production of the area's drylands in a sustainable manner. In any reevaluation of the role of agriculture in the study area, the socioeconomic
impacts as well as the environmental impacts of changing agricultural
practices should be considered.

Prices and Pricing Policies

Policies that subsidize the price of water or emphasize revenue recovery, to the exclusion of considerations of economic efficiency, are especially poorly suited to areas where water is scarce. On the other hand, pricing policies that emphasize economic efficiency and reducing overall water use are appropriate for regions of increasing water scarcity, such as the study area. Appropriate pricing ensures that appropriate signals are sent to consumers about the true cost of water, requiring each consumer to pay the marginal cost of the resources used, and—given some fixed level of benefits—ensures that the costs of providing the water are reduced. Pricing policies that encourage conservation, including marginal cost pricing, time-of-use pricing, and water surcharges generally work best where the quantity of water demanded is reasonably responsive to price.

Augmenting Supplies

Despite the best efforts to reduce water demand through conservation and economic policies, the available freshwater sources in the study area will probably have to be augmented by other sources to meet future needs. This is not at all to say that efforts at reducing demand are futile. Any alternate sources used will be expensive and, in some cases, will furnish lower quality water. Demand management in concert with supply augmentation will be needed to meet the future human and environmental water requirements of the area.

The rainfall of the region is not uniformly distributed over the year—hence there is a premium for storage of runoff when it occurs. In the north, where rainfall is relatively heavy, Lake Kinneret/Lake Tiberias/ Sea of Galilee serves this purpose. In the more arid south, where surface reservoir sites are subject to large water losses from evaporation, subsurface storage has been used extensively. In ancient times, local cisterns were widely used, and this is still a valuable method for developing storage. Artificial recharge of ground water is another method which is currently in use in several places. As urban centers and their paved areas and infrastructure grow, there will be more opportunity to capture runoff from rainfall and use it to recharge ground water.

Additional regional water supplies can be obtained by using what little naturally occurring freshwater is currently unused (through watershed management and water harvesting[1]), by reusing water (wastewater reclamation), by developing sources of lower quality water (use of mar-

---

[1]Watershed management is defined as the art and science of managing the land, vegetation, and water resources of a drainage basin for the control of the quality, quantity, and timing of water, and for the purpose of enhancing and preserving human welfare and nature. Water harvesting is the collection of rainfall by rooftop cisterns.

ginal quality water and desalination of brackish water and seawater), by importing water from outside the study area, by transferring unused water within the study area (water imports and transfers are mentioned but not analyzed in this report), and by attempting to increase the renewable amount of water available (cloud seeding). Again, these options are discussed further in Chapter 5.

Applications and New Research

For each option it is desirable to ask at least two questions. Has an examination been made of all the available information on the option and the factors known to affect their adoption and use? And is it likely that new research might significantly change that assessment? For example, on the demand side no comprehensive study has been carried out of the range of social factors affecting domestic water withdrawal in the study area. At the same time, on the supply side a simpler technology for desalting or filtering water at domestic taps might be developed for arid land conditions. Research agencies have the challenge of deciding what new technology and what mix of technologies and management strategies deserve further exploration. Both kinds of initiatives—canvassing the effectiveness of existing options and exploring innovative technologies—need to be pursued.

Next Steps

This report offers a range of findings and observations on water resource management options. We believe that these options deserve careful examination by the many individuals and organizations who are concerned with the future of water and society in the Middle East. Thoughtful appraisal of experience to date is needed, along with discerning investigation of new relationships and technologies. The results will provide a solid basis for thoughtful, peaceful action to achieve the sustained use of crucial water resources. Rather than suggest a particular political plan, the committee has outlined a broad scope of concepts from which constructive action can emerge.

1

Introduction and Background

We all live on this beautiful water planet which
we have mistakenly chosen to call Earth.

Anonymous

Water—essential for life, economic well-being, and environmental integrity—is in varying degrees in short supply in Israel, Jordan, and the West Bank and Gaza Strip,[1] the Middle East area covered by this report. This region of diverse landscapes experiences both low precipitation and high evaporation. The water scarcity issues for the region's present generation are primarily over distribution of water within the society and preservation of water quality. For future generations, the concerns are to ensure adequate water supplies, preserve the quality of the environment, and achieve greater equity in distributing water throughout the region.

Water supplies adequate to meet basic human needs are clearly essential for maintaining and enhancing the welfare of all inhabitants in the region. Science and technology have critical roles to play in helping to achieve related water resource goals. In focusing on science and related technology, the committee construed its task as consisting of two parts. The first was to identify existing approaches, scientific principles, and knowledge that might be brought to bear on the water problems of the region. The second was to identify opportunities to acquire new scientific

---

[1]In this report, we use the term "the West Bank and Gaza Strip" to refer to Palestinian territory. In the Israeli-Palestinian interim agreement on the West Bank and Gaza Strip, signed September 28, 1995, the term "Palestinian Interim Self-Government Authority" is used to refer to the government. The United Nations General Assembly refers to "Palestine" in its documentation for the governing authority in the West Bank and the Gaza Strip.

knowledge that would be helpful in fashioning sustainable solutions to these problems.

The committee identified five criteria that will be particularly helpful in addressing the region's water problems:

1. Take a regional view. Important insights will be gained by viewing the water problems of the area from a regional perspective, a perspective defined by hydrologic rather than national boundaries. Asking how water quantity and quality problems would be addressed if the region were managed as a single hydrologic unit will yield critical knowledge for good water resource management.

2. Account for the welfare of both present and future generations. The needs of both present and future generations and the status of the environment must all be considered as a matter of equity.

3. Consider all options for balancing water supplies and demands. A perceived gap between estimated future water supplies and water demands is not an adequate basis for water resources planning. Plans must be flexible and robust enough to deal with the uncertainties inherent in hydrologic phenomena, future patterns of social organization and water use, and long-term climatic changes. Plans based solely on projections of expected discrepancies between water supplies and water demands can needlessly reduce the range of planning options to resolve the region's water problems.

4. Maintain ecosystem services to sustain water supplies through integrated planning. Water must be allocated to maintain and enhance environmental quality and biodiversity, in order to sustain water supplies and to preserve the quality of life for the study area's inhabitants.

5. Recognize the mutual dependence of water quality and quantity. Any discussion of the adequacy of water supplies must explicitly acknowledge current and future water quality. The adequacy of water supplies inherently involves issues of water quality. This principle is especially important in the study area, where water is scarce and water quality is deteriorating in many areas.

To apply and implement these critera effectively, it will be necessary to use much of the available scientific and technical information and to continue to seek new scientific information as well. The committee has identified research that would be particularly helpful in resolving regional water problems, such as research on the natural processes that support and deliver ecosystem services. Such promising research and technologies are discussed in Chapters 4 and 5.

WATER AGREEMENTS

On October 16, 1994, Israel and Jordan concluded the Treaty of Peace, which addresses, among other issues, both water and the environment, and establishes a framework for cooperation on water resources. Notably in article 6, Water, the countries recognize their "rightful allocations" of the Jordan and Yarmouk River waters and Araba (Arava) ground water according to the principles and detailed provisions governing quantity, storage, and quality set forth in the treaty's Annex II. In Article 6(4)(d), the two countries agree to cooperate in the "transfer of information and joint research and development in water-related subjects." Annex II establishes a Joint Water Committee to implement the agreement: the countries agree to "exchange relevant data on water resources through the Joint Water Committee" and to "cooperate in developing plans for purposes of increasing water supplies and improving water use efficiency, within the context of bilateral, regional or international cooperation." Annex IV, Environment, obligates the countries to cooperate on Jordan River "ecological rehabilitation, environmental protection of water resources, and nature reserves and protected areas." This commitment underpins our committee's attention to issues of biological diversity and water resources. Appendix A presents the text of the water-related provisions of the Treaty of Peace.

On September 28, 1995, the Israelis and the Palestinians concluded the Interim Agreement on the West Bank and Gaza Strip, which covers water and sewage issues. Article 40 sets forth general principles for water and sewage development, specifies commitments and responsibilities, identifies areas for mutual cooperation, and provides for a joint water committee and joint supervision and enforcement teams (as elaborated in schedules 8 and 9 to the agreement). The text of article 40 and schedules 8 and 9 appear in Appendix B. The earlier 1994 Israel-Palestinian Liberation Organization (PLO) Agreement on the Gaza Strip and Jericho Area also addressed water and sewage issues (article 31) and the protection of nature, nature reserves, and species of special breed (article 23).

On February 13, 1996, in Norway, as an output of the Multilateral Working Group on Water Resources of the Middle East Peace Process, Israel, Jordan, and the PLO signed a Declaration of Principles for Cooperation Among The Core Parties on Water-Related Matters and New and Additional Waters, for the benefit of the Palestianian Authority. This instrument, which specifies recommended voluntary actions, identifies common issues to be included in water resources legislation and management, mechanisms of cooperation on "new and additional water resources," and proposed areas for possible cooperation.

Subsequent to the 1994 Treaty of Peace, the United Nations (UN)

General Assembly approved, on May 21, 1997, a new convention on the Law of the Non-Navigational Uses of International Watercourses, which is now open to signature by states. While this convention explicitly avoids affecting the rights and obligations of state parties to international agreements already in force, it does address water quality and the protection of the ecosystems of international watercourses as important issues in using watercourses. Both these issues are stressed in this report.

Independent of the status of regional peace agreements, the potential for impartial scientific assessments of sustainable regional water supplies provides a continuing opportunity for cooperation.

ROLE OF THE SPONSORING ORGANIZATIONS

In 1994, the presidents and representatives of the Israel Academy of Sciences and Humanities, the Royal Scientific Society of Jordan, the Palestine Health Council, the Egyptian Academy of Scientific Research and Technology, and the U.S. National Academy of Sciences, National Academy of Engineering, and Institute of Medicine met in Washington, D.C., to consider modes of cooperation. They decided there was strong interest in cooperating on joint scientific studies typical of those done at the U.S. National Research Council.

A variety of regional problems and priorities were discussed. High among them was sustainable development in the region. The representatives felt that misuse and overuse of water resources would lead to deterioration of environmental quality, including water quality. Because most Middle East countries share common problems in development, notably in the often related areas of water, environment, and energy, the representatives proposed a study on sustainable regional water supplies.

The following year, 1995, the presidents and representatives of the scientific academies and councils of Israel, Jordan, the Palestinian Authority, and the United States approved a joint study on Sustainable Water Supplies for the Middle East. After consultation among these organizations, a multidisciplinary, multinational committee of volunteers was appointed in December 1995, meeting for the first time in February 1996 and for the fourth and final time in April 1997. Committee members included scientists and engineers from the United States, the Palestinian Authority, Israel, Jordan, and Canada. Together the members have broad expertise, experience, and international perspectives on hydrology, wastewater reuse, water management, surface- and ground-water quality, agricultural and environmental engineering, ecology, biodiversity, natural resources law, agricultural and resource economics, soil and irrigation science, and public health.

The study area charge was to examine ways to increase sustainable water supplies in the Middle East. The study area, as defined in this report, includes Israel, Jordan, and the West Bank and Gaza Strip. The committee was to focus on "methods developed in the Middle East and elsewhere for enhancing water supplies and avoiding over-exploitation of water resources, and on the relationships between water supply enhancement and preservation of environmental quality, especially water quality." Specifically, the committee was to consider "the scientific and technological basis of a range of related issues such as use of treated municipal wastewater for irrigation and other purposes, desalination, water harvesting, cleanup of ground-water contamination, and opportunities offered by improved conservation technologies and strategies to enhance water quality and prevent resource degradation."

Because water resources are critical to both the economic development and the maintenance of natural systems, the committee endorses the concept of intergenerational equity in the use of freshwater resources. This concept is an overarching theme throughout the report. As Edith Brown Weiss has observed, "sustainable development is inherently intergenerational because it implies that we must use our environment in a way that is compatible with maintaining it for future generations." ("Intergenerational Fairness and Water Resources," in NRC, 1993). In many instances, such as withdrawal of ground water in excess of recharge rates or mining of water from nonrechargeable aquifers, there are conflicts between the immediate satisfaction of needs and the long-term maintenance of the resource for both humans and the environment. The 1992 UN Conference on the Environment and Development in Rio de Janeiro endorsed moving away from an emphasis on development of new water supplies toward a focus on comprehensive water management, economic behavior, policies to overcome market and government failures, incentives to provide users with better services, and technologies to increase the efficiency of water use. This recommended focus stresses integrated water management, seeing water not just as a basic human need, but also as an integral part of the ecosystem, a natural resource, and a social and economic good (Box 1.1).

WATER, SOCIOECONOMIC DEVELOPMENT, AND SUSTAINABILITY

As the study area's inhabitants struggle to provide both prosperity for their citizens and good stewardship of the environment, it has been increasingly recognized that economic development, sustainable water supplies, and sustainability are mutually dependent.

Clean and adequate water supplies are critical for long-term economic development; for social welfare, and for ensuring sustainability. Urban

> ### BOX 1.1 Global Water Cooperation
>
> The following principles were endorsed at conferences on water and the environment in Dublin and Rio de Janeiro, both in 1992.
>
> - Water is a scarce resource and should be treated as both a social and an economic good.
> - Water should be managed at the lowest appropriate level, using demand-based approaches and involving stakeholders, particularly women, in decision making.
> - Water should be managed within a comprehensive framework, taking cross-sectoral considerations into account.
>
> The World Bank and UN Development Program have invited other partners to join in establishing a Global Water Partnership to support more coherent and integrated approaches to the management of water resources. A World Water Council has been established at the suggestion of the International Water Resources Association to promote these views.

and industrial growth is creating unprecedented demands for water, often at the expense of agriculture, aquatic ecosystems, and the rural poor (World Bank, 1995). The World Bank suggests that new approaches to water resource management are needed, approaches that will—

- Address quantity and quality issues through an integrated approach
- Link land use management with sustainable water management
- Recognize freshwater, coastal, and marine environments as a management continuum
- Recognize water as an economic good and promote cost-effective interventions
- Support innovative and participatory approaches
- Focus on actions that improve the lives of people and the quality of their environment
- View the management of river basins, coastal zones, and the marine environment together—not as totally separate issues.

The above approach is consistent with many of the conclusions of this committee.

Water and irrigated agriculture have been central in developing the region's economies. The socioeconomic development of both Israel and Jordan since 1950 have depended substantially on the construction of the National Water Carrier and the King Abdullah (East Ghor) Canal. These waterways support production of essential food and fiber, and provide the livelihood for some of the rapidly expanding population of the region. At the same time, the socioeconomic development of the West Bank and Gaza Strip has been hindered by the lack of sufficient water supply owing to both quota restrictions and the absence of infrastructure development (e.g., the proposed West Ghor Canal). Under present conditions, irrigation may continue to play a role in the socioeconomic well-being of the people in the study area. However, new strategies are needed for water use and reuse. Otherwise, severe regional and local water scarcities and contamination will threaten household and industrial sectors, damage the environment, and escalate water-related health problems.

New sources of water are increasingly expensive to develop, limiting the potential for expansion of supplies. As the economies of the region continue to expand, the share of irrigated agriculture in the GNP will diminish. As the human populations grow, the increased demand for potable water will reduce some of the water available to agriculture, but at the same time increase the quantity of wastewater available for irrigation. As irrigation continues in areas overlying aquifers, regional groundwater quality will deteriorate due to increased salinity and other chemical pollutants. These trends will have a profound effect on the role of irrigated agriculture in the area and pose a major water supply challenge.

The primary challenge for regional water resources is to make scarce water as productive as possible, while ensuring equitable distribution of the resource. One important challenge may be reevaluating the role of traditional agriculture. Water used for irrigation will likely have to be diverted to meet the needs of urban areas and industry. New agricultural practices will have to be adopted that are more in harmony with regional climate conditions. Other farming techniques may be considered, such as the "dryland" farming as practiced on the West Bank that requires a minimum of supplemental irrigation, and greater use of hydroponic farms, where water can be recycled to irrigate crops. Brackish water is quite useful for intensive aquaculture in deserts but may increase soil salinity. If efforts are made to improve the productivity of water use in irrigated agriculture and other activities, care should be taken to ensure that gains in productivity are not offset by negative impacts on biodiversity and water quality.

SUSTAINABILITY, INTERGENERATIONAL EQUITY, AND FRESHWATER RESOURCES IN THE MIDDLE EAST

Sustainability

The terms *sustainable* and *sustainability* have been widely used, but they are difficult to define precisely (see, e.g., Norgaard, 1994) and remain the subject of complex interdisciplinary and international research (SCOPE, 1997). When the United States Council on Sustainable Development sought a new consensus for a healthy future society and environment it enumerated ten goals: health, economic prosperity, equity, conservation of nature, stewardship, sustainable communities, civic engagement, population, international responsibility, and education (U.S. President's Council on Sustainable Development, 1996). Sustainable development is defined as the development that meets the needs of the present without compromising the ability of future generations to meet their own needs (UN World Commission on Environment and Development, 1987). Defining any of these goals requires a variety of social and technological assumptions. Nonetheless, the committee agrees that there is an important concept embodied in the terms sustainability and intergenerational equity: the idea that the present generation's children and grandchildren should have at least as much ability to use a resource as the present generation does. This concept should acknowledge the possibility that future generations may use an alternative form of a resource or a substitute. Thus, the committee's interpretation of *sustainability* is incorporated in the term *intergenerational equity*.

Intergenerational Equity

The 1992 UN Conference on Environment and Development in Rio de Janeiro provided a mandate for sustainable development, a goal that requires attention to fairness in using and conserving freshwater resources, both among members of the present generation and between them and future generations. The need for intergenerational equity is particularly acute in the Middle East, where freshwater supplies are scarce, their use intensive, the claimants to them many, and the potential for depriving future generations of adequate freshwater supplies, at least at comparable prices, quite real.

The development and use of freshwater resources raise many intergenerational issues, including potential degradation of resource quality and resource depletion, as well as equitable access to supplies and opportunities to supply human needs. Degraded water quality is perhaps the best known problem. Contamination of surface water, whether in lakes or streams, may cause water to be unusable for some purposes. The natural flushing time for contaminants may be long, and the removal of contaminants expensive. For ground water, the cost of removing contaminants or of containing the spread of toxic contamination may be high enough to preclude most uses of the aquifer. In other cases, pumping ground water in excess of recharge rates may cause saline intrusion into freshwater, which is hard to reverse at acceptable costs and may lead eventually to abandoning parts of the aquifer. These issues also raise important inter- and intragenerational implications. Those who benefit from a lack of concern about contaminating the freshwater sources may not be those who must suffer the consequences.

A second intergenerational issue arises from the depletion of particular freshwater resources. While, physically, water is never lost but only changed in form, the scarcity of freshwater resources in particular places or times may make access to adequate resources more difficult or cause higher real prices to future generations for freshwater.

Equitable access to freshwater resources raises other intra- and intergenerational problems. Poor communities, for example, may suffer serious water pollution, or lack potable water or sufficient water for agriculture and village use—in short, they do not enjoy equitable access to water resources. These conditions may endure from generation to generation.

The report of the Legal Experts for the United Nations Commission on Sustainable Development (UN, 1997a) has identified three principles of intergenerational equity: comparable options, comparable quality, and nondiscriminatory access.

• Ensuring comparable options for freshwater entails maintaining diversity of water supply, from both ground and surface water; developing technology for alternative sources of freshwater at reasonable prices, such as desalination technology; and recycling freshwater resources as possible. Proposed changes to water-intensive agricultural systems might also be reappraised, to the extent that they may deprive others of water for basic needs.

• Ensuring comparable quality means avoiding toxic contamination of watercourses and aquifers, saline intrusion that renders freshwater unusable, and other forms of point and nonpoint water pollution.

• Ensuring nondiscriminatory access for future generations obligates the present generation to attempt to supply water in such a manner that the real price of freshwater is not significantly higher to future generations. In marketing water, price should reflect the full costs, including those to future generations. Moreover, equitable access implies the right to nondiscriminatory bearing of environmental burdens from water pollution and source depletion as well as access to potable water supplies.

The principles of intergenerational equity thus require representation of the interests of future generations, whether in administrative and political decision making, judicial determinations, or the market place.

The actions recommended to the UN Commission on Sustainable Development from the 1997 Assessment of Freshwater Resources of the

World (UN, 1997b) include strategies that address basic human needs and preserve ecosystems consistent with socioeconomic objectives of different societies. Intergenerational fairness implies the necessity for a wide variety of measures, some discussed in this report: good monitoring of water resources quality; scientific and technological research and development for more efficient use of available resources without their contamination or degradation; intergenerational assessments of the effects of particular water projects and uses; effective maintenance of capital investments, such as dams, municipal sewage treatment plants, and water delivery systems; protection of aquifers' recharge areas by appropriate land use planning; and equitable systems for sharing resources among communities. Applying a biogeophysical view, development approaches must be aimed at improving environmental conditions without compromising the capacity to maintain improved conditions indefinitely (Holdren, Daily, and Ehrlich, 1995).

WATER QUALITY, WATER QUANTITY, AND ECOSYSTEM SERVICES

A 1997 UN study, *Comprehensive Assessment of the Freshwater Resources of the World*, reports that consumptive water use has been increasing more than twice as fast as population during this century and that the resulting shortages have been worsened by pollution (UN, 1997b). As a result, one-fifth of the world's people lack access to safe drinking water and more than half lack adequate sanitation. Rapid population increases in the study area—accompanied by intensified industrial, commercial, and residential development—have led to point and nonpoint pollution of surface and ground water by contaminants such as fertilizers, insecticides, human wastes, motor oil, and landfill leachates. Thus, simply maintaining regional water quality and quantity means considering the effects of a wide variety of human activities on watersheds and water bodies.

It is essential to consider the ecological effects and constraints on water development. Water pollution and releases of nutrient-laden municipal sewage effluents have increased, and water consumption has also increased, reducing the flows available for dilution of wastes. Maintaining sufficient freshwater in its natural channels helps keep water quality at levels safe for fish, other aquatic organisms, and people. But regional drainage of wetlands and large-scale ground-water development have had serious negative effects:

- Loss of stream habitat
- Aquatic organisms' becoming extinct or imperiled in increasing numbers

• Impairment of many beneficial water uses, including drinking, swimming, and fishing.

As described in Chapter 4 of this report, biological communities provide many goods and services to humankind: food, fuel, fiber, pharmaceuticals and other biological products, improvement of air and water quality, reduction of climatic extremes, recreation, and aesthetic values. Recent studies are confirming that the capacity of ecosystems to resist changing environmental conditions, as well as to rebound from unusual climatic or biotic events, is related positively to species numbers. The relationship between the natural diversity of biological communities and their provision of goods and services is not precisely known, but it is certain that if diversity is reduced enough, there will be a significant loss of goods and services. One of these services is the provision of clean water.

The benefits, or ecosystem services, provided by freshwater systems fall into three broad categories: (1) the supply of water for drinking, irrigation, and other purposes; (2) the supply of goods other than water, such as fish, vegetation, and waterfowl; and (3) the supply of nonextractive or "instream" benefits, such as recreation, transportation, and flood control. Certain ecosystems, like those found in arid regions, appear particularly vulnerable to human disruption and alteration of their functioning. These sensitive systems all have low representation of key functional types (organisms that share a common role). As society exerts ever greater control and management of ecosystems in the study area, care must be taken to ensure their sustainability, which is in significant part due to the buffering capacity of biodiversity. Consideration of ecological factors related to water is not a luxury, but should be an integral part of planning and management.

THE COMMITTEE'S APPROACH TO THE PRESENT STUDY

The committee's work focuses on scientific and technological information and methods to help maintain or increase water supplies on a sustainable basis in the Middle East. The committee has not considered potential legal constraints, the allocation of water supplies, the desirability of various pricing structures or restrictions on use, international treaties and agreements, or issues of population distribution and growth— except when straightforward descriptions of these policies and their consequences are needed to understand the relevance of scientific and technical discussions.

The committee agreed that this report should accomplish five critical tasks: (1) address options for enhancing water management, without relying on the common or problematic approach of identifying "gaps" between anticipated water supplies and water demand (see Chapter 3); (2) address environmental impacts, including impacts on biodiversity, that are important to consider in planning and managing water resources; (3) analyze water supply options in light of intergenerational equity and sustainable use; (4) discuss a regional approach to water issues, inasmuch as interactions among hydrology, geology, biology, and human populations cross political boundaries within the region; and (5) consider water quality and quantity as closely related.

Estimating of sustainability to achieve intergenerational equity necessarily involves judgments about many parameters. The time horizon is the indefinite future, but forecast accuracy is limited. Climate variability can be estimated from the recent past quite accurately, and at least impressionistically for several thousand years, but future changes cannot be predicted with great confidence, and the potential range of variation is not precisely known. Standards for water quality might change as new insight is gained into factors affecting human health and environmental conditions. The technologies of water extraction, production, treatment, transport, and consumption continue to evolve, with both positive and negative consequences. The number, location, and economic and social status of people in specific areas all shift with growth, migration, and economic development. The committee notes these and other uncertainties where applicable, and describes problem areas where research is needed and likely to be useful in anticipating the likely circumstances of future populations.

The committee has drawn on experience from other parts of the world when it is relevant to Middle East conditions, and it has evaluated to the best of its ability the currently available regional data. The committee has also drawn information and data from recently published reports analyzing regional water resources and demand. Notable examples of these reports include: *Middle East Regional Study on Water Supply and Demand Development* (CES Consulting Engineers and GTZ [Association for Technical Cooperation], 1996), *A Strategy for Managing Water in the Middle East and North Africa* (Berkoff, 1994), and *Core and Periphery* (Biswas et al., 1997). Many other reports that the committee consulted during its study are listed in the bibliography in Appendix E.

This report sets the stage by providing a physical and geographic description of the region and current patterns of water use (Chapter 2). Subjects covered include existing water sources, precipitation, climate, and surface- and ground-water hydrology. Chapter 3 looks at possible future patterns of water use, including the major factors that might influence those patterns. Chapter 4 describes fundamental interactions among the sustainability of water supplies, biodiversity, and ecosystem services in the study area. Chapter 5 presents technological, economic, and other options for improving the region's water supplies, emphasizing integrated water resources management.

The primary audience of this report is scientists and policy makers, especially those of the region; the regional organizations sponsoring this study; nongovernmental organizations; international organizations; and the private and commercial sectors. It has recognized that some future decisions about water will be influenced by policy commitments by the respective countries and by the constraints of national legal systems. The committee was not asked to provide policy advice, and it has not. Rather, the report reflects the committee's consensus on current scientific and technological information about the basic resource and the likely consequences of changes in it. This information will be valuable in other settings to inform good policy choices.

REFERENCES

Berkoff, J. 1994. A strategy for managing water in the Middle East and North Africa. Washington, D.C.: The International Bank for Reconstruction and Development/The World Bank.

Biswas, A. K., J. Kolars, M. Murakami, J. Waterbury, and A. Wolf. 1997. Core and Periphery: A Comprehensive Approach to Middle Eastern Water. Middle East Water Commission. Delhi: Oxford University Press.

Holdren, J., G. Daily, and P. Ehrlich. 1995. The meaning of sustainability: Biogeophysical aspects. In Defining and Measuring Sustainability, M. Munansingha and W. Shearer, eds. Washington, D.C.: The World Bank.

CES Consulting Engineers and GTZ. 1996. Middle East Regional Study on Water Supply and Demand Development, Phase I, Regional overview. Sponsored by the Government of the Federal Republic of Germany for the Multilateral Working Group on Water Resources. Eschborn, Germany: CES Consulting Engineers and Association for Technical Cooperation (GTZ).

National Research Council. 1993. Intergenerational fairness and water resources. Pp. 3-10 in Sustaining Our Water Resources, Proceedings of the Water Science and Technology Board Tenth Anniversary Symposium, November 9, 1992. Washington, D.C.: National Academy Press.

Norgaard, R. 1994. Development Betrayed: The End of Progress and a Coevolutionary Revisioning of the Future. New York, New York: Routledge.

SCOPE. 1997. Sustainability Indicators: A Report on the Project on Indicators of Sustainable Development. Moldan, B., S. Billharz, and R. Matravers (eds.). West Sussex, England: John Wiley & Sons Ltd.

United Nations. 1997a. Report of the Expert Group Meeting on Identification of Principles of International Law for Sustainable Development, Geneva, Switzerland, 26-28 September 1995. United Nations Department for Policy Coordination and Sustainable Development, Background Paper #3, 4th Session UN Commission on Sustainable Development, April-May, 1996. New York, New York: The United Nations.

United Nations. 1997b. Comprehensive Assessment of the Freshwater Resources of the World CSD. New York, New York: The United Nations.

United Nations World Commission on Environment and Development. 1987. Our Common Future. New York, New York: The United Nations.

U.S. President's Council on Sustainable Development. 1996. Sustainable America: A New Consensus for Prosperity, Opportunity, and A Healthy Environment for the Future. Washington, D.C.: U.S. Government Printing Office.

World Bank. 1995. Mainstreaming the Environment. Washington, D.C.: The International Bank for Reconstruction and Development/The World Bank

2

The Study Area and
Patterns of Water Use

The area under study is part of the Middle East, consisting of Israel, Jordan, and the West Bank and Gaza Strip (see Figure 2.1). For simplicity, this area is referred to in this report as "the study area."

Because of various historical and political constraints, data on study area population, and economic and hydrologic features have been difficult to collect and analyze in a common and consistent manner. Many of the quantitative data presented in this report have been contributed by committee members or extracted and modified from a recent report prepared for the Multilateral Working Group on Water Resources and funded by the German Government (CES Consulting Engineers and GTZ [Association for Technical Cooperation], 1996). The German study was prepared by a consulting firm with input from separate Israeli, Jordanian, and Palestinian study teams, and represents the latest and most comprehensive appraisal of the water situation in the study area. While quantitative data from the German study have been used extensively in the present study, the introduction to the German report describes problems in producing study area data because of discrepancies in the interim reports prepared by the individual study teams. No efforts have been made to verify or significantly modify these data for the present report. Use of data from the Multilateral Working Group does not infer endorsement of the validity of that committee's data or findings.

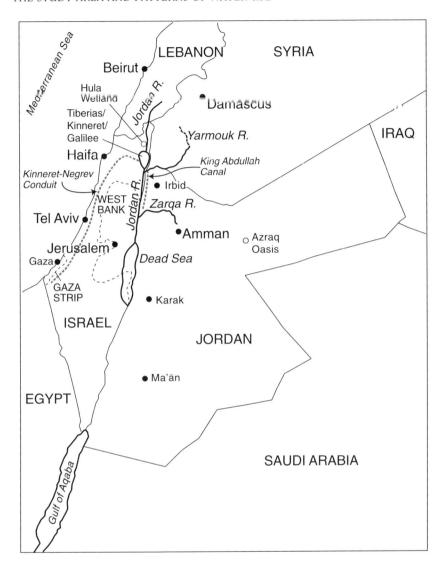

FIGURE 2.1   The study area—Israel, Jordan, the West Bank and Gaza Strip.   SOURCE: Redrawn from published maps.

POPULATION AND ECONOMY

As of 1995, the combined population in the study area was about 12 million (Table 2.1).   Population distribution is highly variable; for the most part, population centers are confined to areas of higher precipita-

TABLE 2.1 Socioeconomic Profile of the Study Area[a]

Characteristics	Israel	West Bank and Gaza Strip	Jordan
Population (in millions)	5.54	2.54	4.33
Literacy (percent)	95	84	87
Gross Domestic Product (GDP)[a]	85.7	2.98	20.9
Composition by Sector (percent)			
Agriculture	3.5	33	6
Industry	22	25	28
Services	74.5	42	66
Per Capita (US$)	16,400	1,300	5,000
Distribution of Workforce (percent)			
Agriculture	3.5	14.1	7.4
Industry	22.1	16.2	11.4
Commerce	13.9	18.2	10.5
Construction	6.5	19.1	10
Transport and Communication	6.3	4.8	8.7
Other Services	47.7	27.6	52

[a]GDP in billion US$; all numbers are for gross comparison only, are based on various sources and reporting years, and may not be fully consistent.

SOURCE: Source of data for Israel and Jordan is the CIA Factbook, 1997 (http://www. odci.gov/cia/publications/factbook/country-frame.html). Source of data for West Bank and Gaza Strip is the Palestinian Central Bureau of Statistics (http://www.pcbs.org/ english/sel_stat.htm) except for data on GDP composition by sector, which are from the CIA Factbook, 1997.

tion. In Israel, more than 90 percent of the population lives in urban localities with the remainder living in rural communities. In Jordan, approximately 78 percent live in urban areas and 22 percent in rural communities. In the West Bank and Gaza Strip, 29 percent live in urban areas, 65 percent live in rural communities, and 6 percent presently (1997) live in refugee camps. The Gaza Strip, where most of the inhabitants live in refugee camps, is one of the most densely populated areas in the world, with an average population per square kilometer of nearly 2,200 people.

An estimated economic profile of the study area is shown in Table 2.1. The higher Israeli per capita gross domestic product (GDP) denotes an advanced economy relative to the remainder of the study area. As will be apparent throughout this report, the effect of this economic difference on the development of water resources is significant and contradictory.

For example, within Israel, the availability of running water to virtually the entire population all the time and the widespread access to water-using appliances have led to relatively high per capita water consumption. On the other hand, the robust Israeli economy has allowed for infrastructure development and agricultural research that have reduced the quantity of potable water used for irrigation per hectare and per production unit. Palestinians and Jordanians, in contrast, with less access to reliable running water and water-using appliances, have lower per capita water consumption but use relatively more potable water for agriculture because of less efficient irrigation systems and a lack of alternative sources (for example, reclaimed wastewater from urban uses).

Agriculture plays a relatively small to moderate economic role in the study area (Table 2.1). However, as will be shown in a section below, agriculture is the principal user of water. In Israel, agriculture accounts for less than 4 percent of the workforce, 3 percent of the gross domestic product (GDP), and 57 percent of the water used (excluding wastewater); in the West Bank and Gaza Strip, 14 percent of the workforce, 33 percent of GDP, and 64 percent of water used; and in Jordan, 7 percent of the workforce, 6 percent of the GDP, and 72 percent of the water used (excluding wastewater). Improvements in irrigation efficiency, increased crop productivity, changes in the types of crops grown (including an increased reliance on dryland farming), and continued growth of wastewater and brackish water irrigation are all clearly required to reduce the freshwater used in agriculture insofar as needed for the growth of other economic sectors and improved living standards and the natural environment.

Economic growth and improved living standards are universal and natural aspirations. Indeed, economic parity and growth throughout the study area are implicit goals of the various multilateral and bilateral peace negotiations. Moves toward parity will tend to increase consumption of water in Jordan and the West Bank and Gaza Strip, thereby placing additional burdens on already-strained water resources. Measures to manage water demand, described later in this report, need to be applied both to lower water consumption in Israel without significantly degrading the standard of living and to minimize increased consumption in Jordan and the West Bank and Gaza Strip while improving the standard of living. If such measures are not taken, economic growth in the study area is likely to be affected by a deficient or expensive water supply.

The study area lacks significant energy resources other than very limited gas fields and undeveloped deposits of oil shale. Electricity is generated almost exclusively with fossil fuels that are imported and therefore costly. Except for the gravity flow of water from the Yarmouk River to the King Abdullah Canal in the Jordan Valley, all water development in the area requires pumping. Because much of the freshwater exists below sea level (in the Upper Jordan River Basin), pumping costs are significant. For example, in Israel, more than 7 percent of the total electricity production is used to transport water. Of the technologies that might be used to augment the supply of water in the study area, some of the more promisingæwastewater reclamation, desalination, and water transfersæare all energy-intensive.

LANDSCAPES

The region has a hot, dry climate, consisting of a dry coast and strip of dry forest upland grading into semidesert and desert. As shown in Figure 2.2, most of the study area is classified as subtropical scrubland, semidesert, and desert. To the north there is an area of temperate steppe and semidesert, and to the south a large expanse of desert, stretching from Egypt into Saudi Arabia. To the southwest the Nile delta abuts, and to the distant east the great neighboring floodplains of the Tigris-Euphrates carry water from Turkey.

The Mediterranean coast is fringed by a gently rolling plain, 15 to 25 kilometers (km) wide, covered by sparse woodland and shrub vegetation. The brown soils of the coastal plain are sandy, requiring large quantities of water and fertilizer to be productive. The plain has a mean altitude of about 50 meters (m) above sea level and gives way to steppe with grassy cover on black and chestnut soils overlying limestone. The steppe gradually rises to form a mountainous upland area with an average altitude of 600 m and having several peaks higher than 1,000 m. The steeper mountain slopes of this upland have been severely eroded and are mostly barren. Soils in the upland limestone areas tend to be shallow, stony, and suitable only for pasture and nonmechanized farming. Alluvial soils in the larger valleys are more suitable for agriculture. The eastern part of the upland is a semidesert with poor soil development.

The transition to the Rift Valley (a part of the Syrian-African rift), east of the uplands is sharp, with land altitudes falling from 600 m above sea level to 200 to 400 m below sea level over a distance of about 15 km. The Rift Valley, a continuation of Africa's Great Rift Valley, contains Lake Kinneret/Lake Tiberias/Sea of Galilee, which is 200 m below sea level, and the Dead Sea, with the world's lowest land altitude of 400 m below sea level. The land parts of the Rift Valley are underlain by fertile alluvial soils, but requiring irrigation to be agriculturally productive.

East of the Jordan Rift Valley is an upland area whose western edge forms an escarpment rising more than 1000 meters above the valley floor and with peaks exceeding 1000 m above sea level. This upland area is 30 to 50 km across and is covered with sparse forest and shrub vegetation. Soil development is generally poor here, except in the northern part, where weathering of basaltic rocks has formed rich, brown soils in some of the larger valleys containing alluvial soils. This upland area grades into a relatively featureless rock- and gravel-covered steppe with alti-

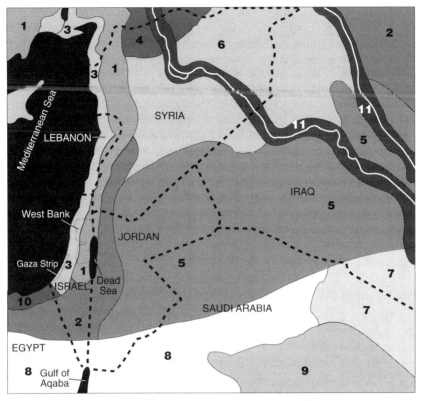

1 Temperate steppe — grasses on chernozem (black) and chestnut soils.

2 Temperate semi-desert — shrubs and small stubby vegetation on brown and gray-brown soils.

3 Subtropical sparse forest — dry summers, thin forests and shrubs on sepia (brown) soils.

4 Subtropical continental — winter transitional steppe; shrubs and small shrub grasslands on gray-sepia soils.

5 Subtropical desert — small shrubs, including succulents and ephemerals on red-brown soils.

6 Subtropical-steppe — small shrubs on gray and sepia soils.

7 Grasses and small shrub grasses on premature soils and sands.

8 Tropical desert — shrubs and small shrubs on premature soils and sands.

9 Tropical desert — devoid of vegetation.

10 Subtropical continental semi-desert — some small shrubs on gray-sepia and gray soils

11 River floodplains and corridors

FIGURE 2.2 Major landscape types in the study area. NOTE: Some scientists would favor using the term "scrubland" for "shrubs." SOURCE: Geographical belts and zonal types of landschafts. Moscow: Principal Organization of Soviet Ministers for Geology and Cartography 1:15,000,000.

tudes ranging from 500 to 700 m above sea level. Vegetation cover is limited to small shrubs. This area is part of the subtropical desert that makes up most of the eastern and southern portions of the study area.

Thus, the natural landscapes of the study area are diverse, although the prevailing visual impression is that of a dry, semiforested coastal highland grading into semidesert and desert. Soil development and natural vegetation, which are largely a factor of climate (see discussion immediately following) are severely limited in much of the area. Naturally arable land is found over 20 percent of the area west of the Jordan Rift Valley and 10 percent east of the valley.

CLIMATE

The study area lies in a transition zone between the hot and arid southern part of West Asia and the relatively cooler and wet northern Mediterranean region. As a result, there is a wide range of spatial and temporal variation in temperature and rainfall. The climate of much of the northwestern part of the area is typically Mediterranean, with mild, rainy winters and hot, dry summers; the eastern and southern parts are much drier, and seasonal temperatures more extreme. Rain occurs in the study area from October to May. Maximum rainfall in the more humid northwestern part occurs in January; the sparse rainfall in the eastern and southern parts is more evenly distributed over the rainy season. Throughout the study area, summers are completely dry, requiring irrigation for crop production.

Mean annual precipitation in the study area is shown in Figure 2.3. Average annual rainfall varies from less than 30 millimeters (mm) in the southern and eastern parts of the study area to locally as much as 1,100 mm in the northwestern part. The average annual rainfall in the central and northern highlands west of the Jordan Rift Valley ranges from 200 to 1,000 mm and ranges in the highlands east of the valley from 200 to 600 mm. Rainfall along the Mediterranean coast ranges from 300 mm in the south to 600 mm in the north. In the southern inland areas, rainfall ranges from 250 mm to 25 mm annually. About 90 percent of the total area east of the Jordan Rift Valley receives less than 200 mm of annual precipitation, and more than 60 percent of the area west of the Rift Valley receives less than 250 mm annually. As is typical of arid and semi-arid climates, there is considerable interannual variability in rainfall.

The variability of precipitation over time is shown in Figure 2.4. This graph, which shows average annual precipitation for the area west of the Jordan Rift Valley, is typical of the entire study area. Precipitation in wet years is almost three times that of dry years. A five- year running mean filters nearly two-thirds of the large year-to-year variation. Nevertheless,

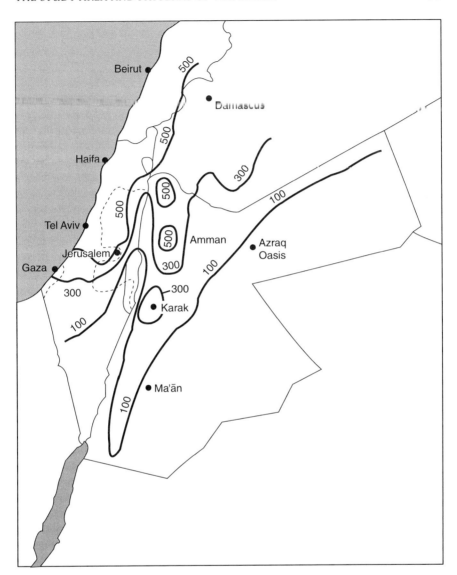

FIGURE 2.3 Mean annual rainfall in the study region in millimeters (mm). Local variations associated with topographic features, including small areas that receive 1,000 mm or more of rain annually, are not mapped. Note that much of the region receives less than 300 mm (11.8 inches) per year. For comparison, mean annual rainfall exceeds 300 mm over about two-thirds of the land area of the United States and over most of Europe. SOURCE: Compiled from information in Salameh and Bannayan, 1993 and U.S. Central Intelligence Agency, 1993.

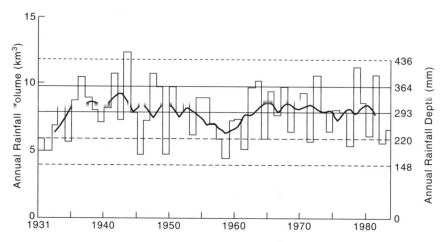

FIGURE 2.4 Annual rainfall in Palestine/Israel, 1931-1984. Solid graphed line connects 5-year running mean. Central horizontal line shows mean value for series; intermediate horizontal line (longer dashes) shows mean ± one standard deviation; top and bottom horizontal lines (shorter dashes) show mean ± two standard deviations. SOURCE: Stanhill and Rapaport, 1988.

because of the lack of significant year-to-year water storage (other than ground water), this variability is a major factor in water-supply planning in the study area.

Summer temperatures are high throughout the study area, generally in the range of 18°C to 32° C, except in the Jordan Valley, where summer temperatures may be as high as 45° C. In the winter, temperatures average about 14°C along the Mediterranean coast and about 9° C at higher altitudes; winter temperatures in the Jordan Valley often exceed 25° C during the day and can be as low as 7° C at night. Temperatures below freezing may occur in upland areas where land elevation exceeds about 500 m above sea level, but they are rare at lower altitudes.

Solar radiation is very high (20 to 30 million joules/m²/day) in the generally cloudless summer months (April through September) and consequently open water evaporation is high in the summer, accounting for as much as 70 percent of the annual total evaporation. Because more water is potentially lost through evaporation than added by precipitation, the study area displays desert characteristics. According to a recent study (Cohen and Stanhill, 1996), solar radiation in the northern Jordan Valley has declined significantly in the past 30 years, possibly owing to greater air pollution. The reduced solar radiation has notably reduced open water evaporation. This observation is potentially significant because, as shown below, the northern Jordan Valley and adjacent highlands are the most important sources of water for the study area.

HYDROLOGY

Precipitation falling on land is either returned directly to the atmosphere by evaporation, flows along the land surface to become surface water, or soaks into the ground. Water that infiltrates into the ground is either drawn into plants and returned to the atmosphere by transpiration or continues infiltrating, becoming ground water. Ground water moves through the earth, eventually re-emerging to the surface in streams, as springs, or as discharge to lakes or the sea. Humans alter the hydrologic cycle by removing surface and ground water, building reservoirs, reintroducing water as sewage discharge to both surface and ground water, and creating additional freshwater through artificial desalination.

Precipitation and Evapotranspiration

According to Stanhill and Rapaport (1988) the volume of annual precipitation in the part of the study area west of the Jordan Rift Valley is about 7,900 million cubic meters (million m^3), based on an average rainfall of 293 mm per year over the 27,011 km^2 area. For the area east of the Rift Valley, the volume of annual precipitation is almost 8,500 million m^3, based on an average rainfall of 94 mm per year over the 89,900 km^2 area (Water Authority of Jordan, 1996 Annual Report). Total precipitation over the study area is therefore about 16,400 million m^3 per year (million m^3/yr). Only a small percentage of this water can be used directly by humans. The high evaporation rates in the study area, combined with transpiration of soil moisture by plants, returns most of the precipitation directly to the atmosphere before it can infiltrate below the soil zone (in "ground-water recharge") or flow directly to wadis, streams, or lakes (as "storm water runoff").

Evaporation rates from open water bodies can be measured. They have been found to range from about 1,550 mm along the Mediterranean coast, to more than 4,400 mm per year in eastern Jordan. These open water, or maximum potential, evaporation rates far exceed precipitation rates shown in Figure 2.3. In assessing water balance, the combined effects of actual evaporation and plant transpiration ("evapotranspiration") from land areas is a more meaningful measurement than open water evaporation. Useful information on evapotranspiration can be derived from measurements or estimates of the other components in the hydrologic cycle (precipitation, ground-water recharge, and stormwater run-off). Such analyses, based on locally measured hydrologic data, indicate that average annual evapotranspiration in the study area ranges from 50 to 100 percent of precipitation.

In some arid parts of the study area, such as the southern and eastern deserts, 100 percent of the precipitation returns to the atmosphere by evaporation and transpiration. It is obvious that in these areas, sustainable water supplies cannot be developed without resorting to engineering schemes, such as artificially recharging storm water before it can evaporate. In other parts of the study area, such as the seasonally humid highlands adjacent to the Jordan Rift Valley, about 70 percent of the precipitation is lost through evapotranspiration. In the sandy, flat, Mediterranean coastal areas, evapotranspiration may be as low as 50 percent of the rainfall. In these areas, sustainable water supplies can be developed by utilizing the precipitation that is not evaporated or transpired. For the area west of the Rift Valley, less than 25 percent of the annual average precipitation is available for human use. East of the Rift Valley, less than 10 percent of the precipitation is available (Al-Weshah, 1992). In the study area as a whole, no more than 17 percent of the 16,400 million m^3/yr of precipitation is available as an average renewable resource (see the summary discussion below on available water in the area). This water is available either as ground water or storm water runoff.

Ground Water

Precipitation that infiltrates into the ground and percolates below the root zone eventually becomes ground water. This process is referred to as "recharge." Recharge rates throughout the study area vary temporally and spatially as a result of variations in the amount and intensity of rainfall and other climatic conditions. Spatial variations are also caused by geological and morphological features of the landscape.

Areas of significant recharge are underlain by permeable and transmissive rocks such as limestone, dolomite, basalt, sandstone, and unconsolidated deposits of sand. These formations that can store and transmit water are known as "aquifers." Aquifers in flat-lying areas generally receive relatively more recharge, because there is less opportunity for stormwater runoff; on some parts of the region's flat coastal plain, almost 50 percent of the rainfall may recharge the underlying sand aquifer. The greatest volume of recharge, however, occurs in the highlands where precipitation is greatest. The permeable limestone and dolomite aquifers that occur beneath the mountainous areas on both sides of the Jordan Rift Valley permit recharge rates of up to 30 percent of precipitation. Recharge to these mountain aquifers accounts for about two-thirds of the recharge in the study area. The areas of highest recharge, and consequently the areas of greatest ground-water availability, are roughly coincident with areas where precipitation exceeds 300 mm/yr.

Total average recharge in the study area is estimated to be 1534 million m^3/yr. About 679 million m^3/yr of this recharge occurs in aquifers shared in unequal areas of land by Israel and the West Bank; 455 million m^3/yr occurs in Israel; 55 million m^3/yr occurs in the Gaza Strip; and 345 million m^3/yr occurs in Jordan. These recharge figures have been adapted from the compilation prepared for the Multilateral Working Group on Water Resources (CES Consulting Engineers and GTZ, 1996, p. 2-6), to provide consistent data across the study area. The data for Israel and the West Bank and Gaza Strip include 220 million m^3/yr of brackish water (Table 2.4 in CES Consulting Engineers and GTZ, 1996). These recharge values are apparently derived from sustainable average annual discharge from the aquifers, with this discharge consisting partially of brackish water pumping (virtually all the water in the Gaza Strip is brackish, because of seawater intrusion and leaching of fertilizers and other salts). Recharge in the Gaza Strip, considering return flows, may be as much as 113 million m^3/yr. Data for Jordan, on the other hand, are apparently based on estimates of freshwater recharge (Table 2.4 in CES Consulting Engineers and GTZ, 1996). Because parts of some of the country's aquifers, such as in the Jordan Valley, also contain brackish water from agricultural activities and the upward movement of deeper saline water, 70 million m^3/yr of brackish ground water (Table 5.2 in CES Consulting Engineers and GTZ, 1996) have been added to the Working Group's recharge value for Jordan to derive a figure comparable with the Israeli and Palestinian figures. The committee is aware that there are a variety of estimates differing from these in various respects, but uses these as conveying a rough picture of the situation.

Ground water moves from areas of recharge to areas of discharge. Natural ground-water discharge generally occurs in low-lying areas, such as the coast, the Rift Valley, and interior depressions. Ground-water discharge includes the base flow or fair weather flow of streams, the flow from springs, and seepage to surface water bodies, such as the Mediterranean Sea, Lake Kinneret/Lake Tiberias/Sea of Galilee, and the Dead Sea.

Under natural conditions, aquifers are in a state of dynamic balance. During periods of recharge (October to May), water is added to storage within the aquifer, and the water level or water pressure within the aquifer increases, causing natural discharge to increase. During periods of no recharge, water levels decline and natural discharge is diminished or, in some cases, ceases. Over the long term, and in the absence of human intervention, natural recharge equals natural discharge; the amount of ground water stored in the aquifers, although constantly fluctuating, remains about the same.

Humans have greatly altered this natural state of balance by pumping water from wells. Withdrawals from wells are referred to as "pumping" even though on occasion, as in parts of the Yarmouk Basin Aquifer, artesian pressure may be sufficient to cause wells to flow without the use of pumps. The specific yield of an aquifer declines as more water is pumped from it and increases as more water is stored: the ground water pumped from wells is balanced by a combination of increased recharge, decreased natural discharge, or decreased storage in the aquifer up to the point where diminished storage forces a decrease in withdrawal rate. Increased recharge occurs when shallow saturated rocks are dewatered so that the aquifer can accept and store more water (although in some cases the substrate can consolidate and reduce storage capacity). But this phenomenon is of negligible importance in the overall water balance of the study area.

Decreased natural ground-water discharge as a result of pumping is well documented in the study area. The disappearance of natural spring flow in the Azraq Basin and the drying up of the springs that provided perennial flow to the Yarkon River have each been affected by ground-water pumping. Because of this relationship between ground and surface water, recharge values should not necessarily be considered as independent sources of water. For example, about 175 million m³/yr of ground water discharges into Lake Kinneret/Lake Tiberias/Sea of Galilee (Table 2.2 in CES Consulting Engineers, 1996) and significant development of the aquifers adjacent to this lake would decrease this discharge, affecting the amount of surface water available. Although the relationship of ground and surface water is generally well understood, it is not clear how it is accounted for in water management and planning in the study area.

If long-term rates of withdrawal do not exceed the rate of recharge to an aquifer, the withdrawals will mostly be balanced by decreased natural discharge with only minor quantities of water withdrawn from storage. If withdrawals exceed recharge, storage within the aquifer will be depleted. Figures 2.5 and 2.6 depict water level fluctuations for two wells in the study area—one an aquifer where long-term storage is constant, and the second an aquifer in which storage is being continually depleted. The first case (Figure 2.5), illustrates a recurring annual cycle, in which water levels decline during the nonrecharge period as water is withdrawn from storage by means of both natural and artificial discharges. Because the average rate of withdrawal does not exceed the average rate of recharge, water levels increase during periods of recharge, and the water withdrawn from storage is replaced. In the second case (Figure 2.6), although water levels increase during the recharge period, the quantity of water

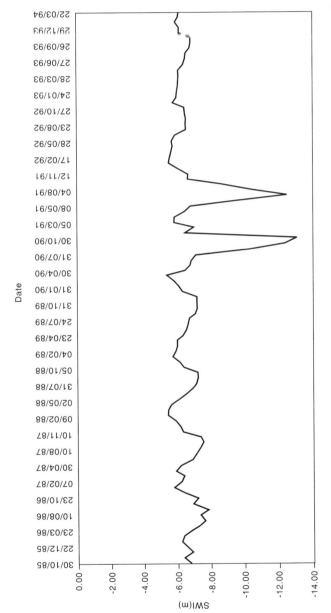

FIGURE 2.5 Water fluctuations in a well with stable level of storage. SOURCE: Water Authority of Jordan, open files.

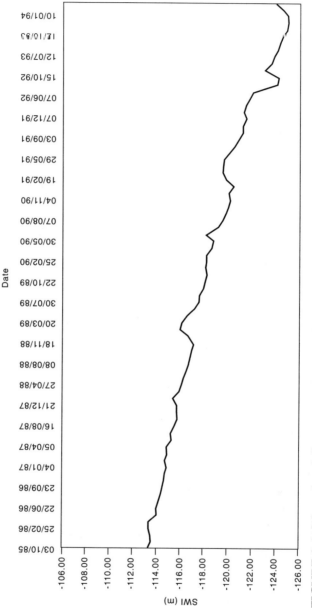

FIGURE 2.6 Water level fluctuations in a well that is being depleted. SOURCE: Water Authority of Jordan, open files.

added to the aquifer is not enough to completely balance the amount discharged. The overall trend of the water level is downward, representing a continuing depletion of storage. Similar records are available for Israel and the West Bank and Gaza Strip.

As illustrated in Figures 2.5 and 2.6, the aquifers in the study area serve as reservoirs, allowing ground water to be used throughout the year and providing for drought protection. The total volume of freshwater contained in the area's aquifers is several billions of cubic meters. Although the depletion of ground-water storage has been only a small percentage of the total storage, it is indicative of an inherently unsustainable water supply. Ground-water pumping in the Gaza Strip, for example, exceeds recharge by about 2 to 18 million m^3/yr; this excess pumping represents water that is permanently removed from storage in the underlying aquifer each year.

Over the past 10 years, the Disi aquifer in Jordan has been pumped at rates far exceeding its recharge rate, if it receives any recharge at all, resulting in a loss of storage expressed as a lowering of water levels by as much as 80 m. If rates of pumping continue to exceed the recharge rate, water level in the aquifer will ultimately decline below the depth at which they can be economically extracted. However, one justification for overpumping some aquifers for a short time is that it may permit development of a nonagricultural economy in a local area while transition is made to a genuinely sustainable economy in the future. Meanwhile, a nonsustainable agriculture is continued temporarily.

Of equal importance to water planners is the issue of drought contingencies. Although a drought year may severely affect the availability of surface water, it only affects the availability of ground water slightly, because of storage within the aquifers. Ground water is therefore likely to be overpumped during droughts to compensate for surface water deficiency. If the drought lasts for only one or two years, subsequent wetter years may largely replenish the losses in ground-water storage that result from the overpumping and diminished recharge during the drought. A series of drought years, however, may eventually lead to serious depletion of ground-water storage, with consequences for both short- and long-term water supply. Because of the uncertainty of future droughts, and the unknown effect on water supply caused by possible climate change, it is important to consider the probability and effect of drought years. Techniques such as optimization and simulation modeling can be used to prioritize the allocation of sustainable water resources for a variety of drought scenarios.

Surface Water

Estimated average annual surface water flow in the study area is about 1,429 million m³/yr. This figure is based on an average annual estimated flow of 1,300 million m³/yr in the Jordan Basin (CES Consulting Engineers and GTZ, 1996, p. 2-3), 47 million m³/yr in interior basins in Jordan (Salameh and Bannayan, 1993), 80 million m³/yr in coastal basins in Israel (Table 5.1 in CES Consulting Engineers and GTZ, 1996), and 2 million/m³/yr in the Gaza Strip (Palestinian Water Authority, written communication). Rates of stream flow in the study area vary significantly from year to year; in drought years, annual stream flow in the Jordan Basin may be as low as 420 million m³/yr and during wet years it may be as high as 2,460 million m³/yr (Table 2.2 in CES Consulting Engineers, 1996). Within-year fluctuations are also pronounced, with highest flows coinciding with the October to May rainy season.

Streamflow has two components: base flow and stormwater runoff. Ground water that discharges as springs or that seeps directly into streambeds constitutes the base flow or "fair weather" portion of stream flow, and accounts for virtually all the flow during the annual dry season. Perennial streams, that is, streams with a base flow component, exist only in the northwestern part of the study area, where ground-water recharge rates, and consequently ground-water discharge rates, are highest. These perennial streams include the Yarmouk River and the Upper Jordan River and its tributaries, such as the Dan River. As noted earlier, some formerly perennial streams, such as the Yarkon River on the coastal plain, no longer flow during the summer, owing to decreased natural ground-water discharge from the pumping of wells. This elimination of spring flow was planned, as part of a strategy to develop the water resource efficiently. However, instream use of perennial streams by native flora and fauna is an important and often overlooked benefit; the loss or reduction of perennial streamflow has had a profound effect on local biodiversity.

Because of the sensitivity of the base flow component of streamflow to ground-water withdrawals, surface water flow (1,429 million m³/yr) and ground-water recharge rate (1,534 million m³/yr) are not additive. This fact should be a significant consideration in water planning, particularly in the northwestern watersheds, which are the source of water transferred to the more arid southern areas via the National Water Carrier and the King Abdullah Canal. Surface- and ground-water supplies must be developed together, and not as separate resources. Unfortunately, studies to quantify the base flow component of streamflow in a consistent manner throughout the study area are lacking. As a result, efforts to define the availability of water accurately and to predict the consequences of withdrawals are somewhat hampered.

Stormwater runoff, or flood flow, is the other source of water in streams. The streams draining the northwestern part of the study area, where precipitation exceeds 500 mm/yr carry most of the annual stormwater runoff. Because the frequency, magnitude, and distribution of storm runoff are highly variable, and because many of the wadis tributary to the Rift Valley and elsewhere in the study area are not gauged, the quantity of stormwater runoff is not precisely known.

Except for very wet years, about half of the runoff in the relatively humid northwestern part of the study area is held behind impoundments and eventually utilized for water supplies. An estimated 100 million m^3/yr of runoff occurs in unregulated streams and wadis. Additional development of this stormwater runoff is discussed in the section on Water Harvesting in Chapter 5. Unlike base flow, stormwater runoff has probably increased through human activities, because runoff rates are generally higher from paved areas and rooftops than from natural surfaces. The concentration of storm water in urban areas provides a particularly favorable opportunity for the development of additional sources of water.

Some floodwaters naturally recharge underlying aquifers, particularly in the wadis tributary to the Rift Valley, and so they cannot be considered a source of supply completely above and beyond the associated ground water. More significantly, the widespread occurrence of storm water in hundreds of small streams and wadis, together with the variability of storm water from year to year, makes it impractical to construct impoundments that can trap all the flow. Additionally, elimination of storm flow will have adverse ecological effects on downstream waters and adjacent ecosystems.

Because of the temporal variability of surface water, this water source requires storage to increase its usefulness as a water supply. Lake Kinneret/Lake Tiberias/Sea of Galilee, a natural lake, is now highly managed to utilize the flows of the upper Jordan River drainage basin. Also, several dams have been constructed on west-flowing tributaries of the lower Jordan River, the largest of which is the King Talal Dam on the Zarqa River. The Yarmouk River, although not impounded, is diverted, partly into Lake Kinneret/Lake Tiberias/Sea of Galilee and partly into the King Abdullah Canal. Surface water management of the Jordan River watershed through impoundments and diversions provides almost one third of the water used in the study area. As a result of this management, average flow of the lower Jordan River has decreased from almost 1,400 million m^3/yr to less than 300 million m^3/yr (Salameh and Bannayan, 1993). The water remaining in the river is return flow from irrigated fields, the discharge of saline springs, and in wetter years, storm water runoff.

On one hand, the decrease in the flow of the lower Jordan River can be regarded as a positive result of upstream water supply development, including 620 million m^3/yr contributed to the National Water Carrier and 120 million m^3/yr to the King Abdullah Canal. Additional water is also withdrawn from the Yarmouk River in Syria. On the other hand, the decreased flow and the poor quality of the remaining flow have had strongly negative effects on natural flora and fauna in the lower Jordan River, and have lowered water levels in the Dead Sea.

Other Sources of Water

Humans have greatly affected the natural hydrologic cycle in other ways. Withdrawal of water has altered streamflow and spring discharge and impaired water quality. A unique feature of the study area is the degree to which used water ("wastewater") is recycled and used again. Water supplies in the study area are supplemented by the use of 272 million m^3/yr of reclaimed urban wastewater in the agricultural sector. A major concern in reusing wastewater is, of course, water quality. The present and potential use of wastewater as a water supply is discussed in the section on wastewater reclamation in Chapter 5.

In some cases, aquifers below the zone of active ground-water circulation may contain freshwater in what are referred to as "fossil" or "non-renewable" aquifers. Because these aquifers receive little or no recharge, they cannot be said to have a sustainable yield. Nevertheless, they hold billions of cubic meters of usable water. It is estimated that about 95 million m^3/yr of nonrenewable water are withdrawn from fossil aquifers in the study area. The consequences of this approach are discussed under the section on ground-water overdraft in Chapter 5.

Human activity can also create "new" freshwater through artificial desalination of water with excess salinity. This source is currently negligible within the study area, but its potential is discussed in a following section in Chapter 5 on desalination of brackish water. Another source of water is importation from outside the study area. This practice is not presently used in the study area, but some potential approaches are briefly described again in Chapter 5, in the section on imports of freshwater into the study area. Water importation has also been the object of analysis in recent planning reports, such as Biswas et al. (1997).

Water Quality and Salinity

One requirement of a sustainable water supply is that the biological, chemical, and physical characteristics of this water be suitable for its intended use. Characteristics that render a water source unsuitable for a use may be natural, such as high salinity in some springs, or they may be introduced by human activities, such as toxic chemicals discharged to a natural water body. Treatment can, of course, turn any source into a suitable water supply, with desalination of seawater being an extreme example. Polluters rarely pay the cost of cleanup; rather, these costs usually reflect inefficiency and are borne by downstream users, or, as is commonly the case with ground-water pollution, by a future generation of users. The costs of treating water impaired by human activity therefore raise issues of both economic efficiency and intra- and intergenerational equity.

The chemical quality of ground water is determined by the nature of the rocks through which it moves. Some aquifers are composed of rocks, such as limestone, that are easily dissolved. Ground water moving through them, then, contains relatively large amounts of dissolved solids. Other aquifers are composed of relatively insoluble rocks, such as sand or sandstone, and these contain water with low dissolved solids. Human activities in recharge areas also greatly affect ground-water quality. Fertilizers, insecticides, leachate from landfills, and domestic and industrial wastes, for example, can contaminate aquifers.

Ground-water contamination by human activities at or near the land surface has only been recognized in recent years. Before about 1980, in some quarters it was believed that soils served as filters, preventing harmful substances deposited at the surface from migrating downward into ground water. Today it is known that soils and other intervening layers have a finite capacity to filter and retard substances and to protect aquifers from contamination (NRC, 1993). Because of the slow rates, and sometimes circuitous directions, of ground-water movement, remedial actions may be difficult or impossible. Protection of ground-water quality is therefore a prime consideration in the sustainability of the water supply.

Aquifers below the circulation zone generally contain brackish or saline water. Aquifers adjacent to seawater also contain saline and brackish water. Withdrawal of fresh ground water can induce the movement of brackish or saline water into freshwater parts of aquifers. Saltwater contamination is difficult to reverse and therefore also affects sustainability.

The chemical quality of surface water is determined by the quality of its two components: base flow and stormwater runoff. Ground water discharging into streams is at the end of a sometimes lengthy flow path and may contain relatively high concentrations of dissolved solids. The material dissolved in base flow generally consists of natural substances derived from the rocks that make up the aquifer, such as calcium, magnesium, sulfate, bicarbonate, and iron. Stormwater runoff generally contains low concentrations of these substances, because it has limited contact with rocks containing them. Storm water, however, often contains a high sediment load and, particularly when generated over urban or agricultural areas, may contain high concentrations of nitrates, phosphorus and other nutrients, as well as metals and organic material from fertilizers, pesticides, animal wastes, and trash.

Surface water generally requires more treatment prior to human consumption than does ground water. However, in the natural environment, plants and animals are in contact with untreated surface water, which often contains urban, agricultural, or industrial contaminants. Even relatively low concentrations of some of these contaminants in surface water can stress the biological integrity of lakes, reservoirs, and other receiving waters through various physical and chemical processes. Metals can be toxic to aquatic life and are often accumulated and concentrated in the food chain. Excess nutrients commonly result in the growth of undesirable aquatic plants and contribute to eutrophication. Suspended material in water causes turbidity, which affects light penetration and inhibits the growth of desirable aquatic plants. Biodegradable organic material (as measured by biochemical and chemical oxygen demand) can reduce the level of oxygen dissolved in water, killing fish.

An intrinsic problem of water management in arid areas is salinity. Salts are continually added to soils through water dissolution of rocks and minerals and degradation by plant roots (for example, the salt calcium carbonate), input of aerosols and recharge (for example, the salt sodium chloride), as well as by anthropogenic activity (for example, various salts of nitrogen and phosphorus as a result of fertilizer use). Where rainfall is sufficient, such salination is balanced by salts leaching out of the soil by infiltrating water (recharge). The salts are incorporated in groundwater and eventually discharged to streams or directly to the ocean, seas, or other terminal discharge points. In this way, humid areas are net exporters of salts and salt accumulation in soils is virtually nonexistent.

Arid regions, on the other hand, are generally net importers of salts, and salt accumulates in soils. Within the study area, experience has shown that salt accumulates in soils where annual precipitation is less than about 300 mm. Where precipitation is higher than about 400 mm, winter rains are effective in leaching out salts that have accumulated in the root zone during the previous summer. From the mean annual precipitation shown in Figure 2.3, it can be seen that only a very small part of the study area is able to sustain long-term agriculture without a management scheme that artificially removes salt accumulated in the soil.

Even within the humid parts of the study area where precipitation exceeds 400 mm/yr, the natural seasonal balance of salt accumulation and rleaching has been disturbed by human activity. Unlike precipitation, which contains very low concentrations of salts, irrigation with surface and ground water adds considerable salts to the soil zone. Even more salts are added when the source of irrigation water is recycled agricultural and municipal wastewater.

To recycle as much water as possible, drainage from agricultural lands is often recaptured and reused for irrigation. The use of this water, enriched in salts, can lead to serious problems of soil salination, as has occurred in the J'esreel Valley in Israel after about 20 years of such practice. Reuse of municipal wastewater is another serious source of salinity. Through water softening, food processing, use of detergents, and physiological sources, urban wastewater is enriched with salts. For example, an excess of 100 to 200 milligrams per litter (mg/l) of chloride is contained in wastewater in Israel when compared to the source water. Most treated wastewater in the study area has a dissolved solids content of more than 1,000 mg/l. The use of such water for irrigation has led to crop damage and reduced yields and may eventually lead to irreparable soil salination.

Both percolation of saline irrigation water and the leaching of salt by natural recharge lead to increases in dissolved solids in underlying ground water. These trends have occurred along the Mediterranean coast in Israel and the Gaza Strip, in the Jordan Valley and the Dhuleil area east of Amman, and elsewhere. The withdrawal of this enriched water for irrigation or domestic use (with ensuing treatment and reuse for irrigation) creates a cycle of increasing salination of both soil and ground water. Furthermore, the almost total elimination of natural ground water and surface water discharge effectively blocks the natural discharge of salts, so that the study area is now importing salts but not exporting them. The salts leached by winter rains from irrigated soils above the coastal aquifer accumulate in ground water at a rate of 2 mg/l chloride per year. While chloride concentration in ground water was 110 mg/l in 1963, it was 170 mg/l in 1993. Similar increases in chlorine contamination have been seen elsewhere in the study area.

One lesson from the area's history of soil and ground water salination is that total utilization of water, when coupled with significant internal recycling of water through reuse, is not sustainable because of progressive salt accumulation. A sustainable water management system should include salt removal, by natural leaching, removal of salty drainage water, or desalination.

Available Water in the Area

In the context of sustainability, the simplest definition of "available" water is the average quantity of water available on a renewable basis. The

TABLE 2.2 Estimated Annual Average Renewable Surface- and
Ground-Water Resources (in million m³/yr)

Source of Data	Surface Water Resources	Ground Water Resources	Total Renewable Resources
Multilateral Working Group	1,200	1,400	2,600
This NRC Committee Report	1,429	1,359	2,788

SOURCE: Multilateral Working Group data from CES Consulting Engineers and GTZ,
1996, p. S-3.

average annual rates of ground-water recharge (1,534 million m³/yr) and
surface water flow (1,429 million m³/yr) are renewable resources. Unfortunately, these two water sources together cannot provide the total annual renewable water resources of the study area. As already noted,
water moves from aquifers to streams (as base flow) and also from streams
to aquifers (as recharge of storm water runoff). An apparent attempt to
reconcile all these factors has been made by the Multilateral Working
Group on Water Resources (CES Consulting Engineers and GTZ, 1996).
Although their definitions and methods are unclear, they estimate that
annual average renewable water resources in the study area are 2,600
million m³/yr, consisting of 1,400 million m³/yr of ground water and
1,200 million m³/yr of surface water (Table 2.2). It is not clear how the 175
million m³/yr of ground water that discharge into the upper Jordan River
and Lake Kinneret/Lake Tiberias/Sea of Galilee (Table 2.2 in CES Engineers and GTZ, 1996) are accounted for. An alternate calculation of renewable water resources, recognizing that there may be differences in
accuracy of field data, would be to subtract the 175 million m³/yr from
the total annual ground-water recharge (estimated in this report to be
1,534 million m³/yr), because it is eventually accounted for (and used for
water supply) as surface water. This calculation yields an annual average
renewable water resource of 2,788 million m³/yr (Table 2.2). Development of the total renewable resource, whether 2,600 million m³/yr or
2,788 million m³/yr, would be highly impractical because of the difficulties inherent in capturing all the storm flows.
    The large amount of water stored in aquifers that could provide sup-
plies in excess of natural replenishment is not considered in the above
estimates of available water resources. Although short-term use of such
storage to moderate the temporal variation in recharge is a standard water management practice, long-term and continual use of storage, or
ground-water mining, would seem to be inherently a nonsustainable
development. Use of fossil ground water is also nonsustainable, and this

resource is therefore not included in estimates of renewable water as well. However, it is estimated that 95 million m^3/yr of nonrenewable fossil water is being used annually in the study area (25 million m^3/yr in Israel and 70 million m^3/yr in Jordan), and the Multilateral Working Group on Water Resources (CES Consulting Engineers and GTZ, 1996, p. 2-7) includes as much as 253 million m^3/yr of fossil water (110 million m^3/yr west of the Rift Valley, and 143 million m^3/yr east of it) in their future water supply scenario. Social and economic issues of intergenerational equity, brought about, for example, by increased water use in new economic development, clearly need to be addressed before the fossil supplies are depleted. Hydrologic issues concerning the amount of stored ground water that can be pumped and other consequences of that pumping also need to be addressed.

Additional brackish ground water and nonconventional sources of water, such as wastewater reuse, desalination, and imports, are also available in addition to the renewable water resources. Although about 290 million m^3/yr of brackish water are included in the renewable sources, hydrologic studies are needed to determine whether this or greater quantities of brackish water can be developed without significant deterioration of resource quality. The use of brackish water, primarily for industry and agriculture, is further addressed below in the section on use of water of marginal quality in Chapter 5.

At the present time, only wastewater is a significant nonconventional source of water in the study area. About 309 million m^3/yr of wastewater are currently used in the area, 250 million m^3/yr in Israel and 59 million m^3/yr in Jordan. Opportunities to use greater amounts of wastewater are discussed in a following section on wastewater reclamation in Chapter 5. As much as 1,794 million m^3/yr of area wastewater may be available for reuse in the future (Tables 5.1 and 5.2 in CES Consulting Engineers, 1996). Such large-scale use of recycled wastewater may lead to significant salination of soils and groundwater, as previously described.

Efficient use of the available water requires a high degree of planning and management, which is made more difficult by political considerations. For example, agreements between Israel and Jordan on the allocation of water from the lower Yarmouk River (Appendix A) are based on anticipated conditions of flow in the river. Average flow conditions, however, have changed in the past and may change again in the future: average annual flow between 1927 and 1964 was 467 million m^3/yr; between 1950 and 1976, 400 million m^3/yr; and for the most recent period about 360 million m^3/yr (Salameh, 1996, p.16). In addition to possible climatic changes, decreased flow in the Yarmouk River is attributable to upstream riparian use of surface water as well as groundwater withdrawals. Because much of this upstream use is in Syria, which is not a party to the

Israel-Jordan agreements, future availability of water from the Yarmouk is uncertain. In general, the use of transboundary water sources is subject to political considerations, but hydrologic and environmental considerations are more likely to lead to sustainable use (e.g., Feitelson and Haddad, 1995).

Another example of hydrologic uncertainty owing to political considerations is presented by the so-called mountain aquifer, underlying both Israel and the West Bank. Indiscriminate ground-water development in such cases, on either side of any future political boundaries, may result in significant depletion of this resource. To maximize the sustainable yield of this aquifer, the location of wells must be determined by hydrologic, not political, considerations. Options for the joint management of aquifers shared by Israelis and Palestinians are presented in Feitelson and Haddad (1995).

Planning and management of surface and ground water are also made difficult by inconsistent data. As pointed out earlier, it has been difficult to obtain hydrologic data collected and analyzed in a common and consistent manner, or other such study area data for hydrologic syntheses (CES Consulting Engineers and GTZ, 1996). Allocation of surface water (for example, from the Yarmouk River) or development of a shared aquifer (for example, the mountain aquifer) will be facilitated if all parties use common methods to collect and analyze pertinent data, and if all parties have access to these data. In recognition of this, the Multilateral Working Group on Water Resources implemented a Water Data Banks Plan in 1995 to work toward common methods of collecting water data in the study area. Also needed, however, are joint study teams to provide regionally consistent meteorological and hydrological analyses, if the water resources of the study area and the consequences of various water development scenarios are to be fully assessed.

WATER USE

The total water used in the study area in 1994 was estimated to be 3,183 million m^3/yr. Table 2.3 shows the breakdown by type of use (domestic, agricultural, and industrial) and by water source (ground water, surface water, and treated wastewater). From these data it can be seen that agriculture is the principal user of water, accounting for 66 percent of total use; about 62 percent in Israel, 64 percent in the West Bank and Gaza Strip, and 73 percent in Jordan. Excluding wastewater, agriculture accounts for 63 percent of the total surface- and ground-water withdrawals in the study area, 58 percent of withdrawals west of the Rift Valley and 72 percent east of the valley. Efforts to reduce the amount of freshwater used in the agricultural sector have been an important part of

TABLE 2.3 Estimated 1994 Water Use in the Study Area, by Subarea, Type of Use, and Water Source (in million m^3/yr, except per capita use in m^3/yr)

Type of Use	Israel	West Bank and Gaza Strip	Jordan	Total
Domestic				
Ground Water	—[a]	85	208	—[a]
Surface Water	—[a]	0	33	—[a]
Wastewater	0	0	0	0
Subtotal	545	85	241	859[b]
Agriculture				
Ground Water	—[a]	150	331	—[a]
Surface Water	—[a]	0	382	—[a]
Wastewater	213	0	59	272
Subtotal	1,180	150	772	2,102
Industry				
Ground Water	—[a]	0	43	—[a]
Surface Water	—[a]	0	0	—[a]
Wastewater	0	0	0	0
Subtotal	129	0	43	172
Conveyance Losses[c]	50	—	—	50
Total Water Use				
Ground Water	1,006	235	582	1,811[b]
Surface Water	685	0	415	1,100
Wastewater	213	0	59	272
Total	1,904	235	1,056	3,183[b]
Gross Water Use Per Capita	344	93	244	257

[a]Because both ground water and surface water are conveyed in the Israeli National Water Carrier and distributed to domestic, agricultural, and industrial users, the precise sources of these supplies cannot be determined.

[b]This figure is adjusted to eliminate the double counting of about 7 million m^3/yr supplied to East Jerusalem and 5 million m^3/yr supplied to the Gaza Strip by Israel and included in both the Israeli and the West Bank and Gaza Strip totals.

[c]Conveyance losses are reported only for the National Water Carrier in Israel. Other service distribution systems, such as the King Abdullah Canal in Jordan and all piped distribution systems, have losses that are included in the water-use figures.

SOURCE: Adapted from CES Consulting Engineers and GTZ, 1966.

water planning in the area, and have included, in addition to conservation measures, increased use of treated wastewater and brackish ground water. As a result of the increasing use of treated wastewater as a source of irrigation supply, the percentage of surface- and ground-water withdrawals used by agriculture has decreased from an estimated 80 percent in 1965 to the current 63 percent (Table 2.3). Further reductions in agricultural water use are discussed in the section on options for the future in Chapter 5.

There is significant disparity in water use among Israel, the West Bank and Gaza Strip, and Jordan, particularly in per capita domestic water use. Average per capita use of water in the domestic sector, based on population data from Table 2.1 and use data from Table 2.3, is 98 m^3/yr in Israel, 34 m^3/yr in the West Bank and Gaza Strip, and 56 m^3/yr in Jordan. These per capita figures reflect water put into urban distribution networks; as much as half this water may be lost due to leaky pipes in some systems. As previously discussed, moves toward economic parity will tend to increase water consumption; one planning scenario projects future per capita domestic use of water at 90 m^3/yr in Israel and 70 m^3/yr in Jordan and the West Bank and Gaza Strip (CES Consulting Engineers and GTZ, 1996, p. 3-11).

The heavy reliance on ground water in the study area is apparent from Table 2.3. Ground water accounts for 57 percent of total water used and 62 percent of water withdrawn. Current total ground-water withdrawals of 1,811 million m^3/yr (Table 2.3) are in excess of the estimated 1,359 to 1,400 million m^3/yr of renewable ground-water resources (Table 2.2). The "overpumping" of between 411 and 452 million m^3/yr consists of 95 million m^3/yr of fresh fossil water (25 million m^3/yr in Israel and 70 million m^3/yr in Jordan), 81 million m^3/yr of brackish fossil water in Israel (Table 2.4 in CES Consulting Engineers and GTZ, 1996), with the remainder, about 250 million m^3/yr, taken from storage, most of it from aquifers in Jordan. The total overpumping in Jordan may be as much as 307 million m^3/yr of renewable water.

In contrast to ground water, Tables 2.2 and 2.3 together indicate a surplus of between 100 and 329 million m^3/yr of surface water. This surplus largely represents uncaptured storm water runoff in the wadis tributary to the Jordan Rift Valley. According to various studies, more than 200 million m^3/yr of storm water runoff may currently be captured by retention structures throughout the study area (BRL-ANTEA, 1995; and Water Authority of Jordan 1996, open files). The possible use of storm water to augment existing water supplies is discussed under watershed management in Chapter 5.

THE IMPORTANCE OF HYDROLOGIC RELATIONSHIPS
IN THE STUDY AREA

In reviewing information on the study area, the committee was struck by the absence of a consistent, comprehensive, and reliable body of data on the availability and use of water resources. This lack is attributable to three different problems. First, the reliability of some data is open to question. There is no agreement, for example, about the quantities of surface water available for diversion in an average year. Second, some important measurements are either incomplete or absent altogether. For example, there is no comprehensive characterization of the temporal variation in precipitation and runoff. Finally, data are sometimes gathered on different and incommensurate bases, leading to difficulties in characterizing the water-related features of the study area.

The committee notes that, regardless of national boundaries, the waters of the area are shared because the region is hydrologically connected. Changes in the quantities and qualities of water available in one area will have impacts on the quantities and qualities available in other areas. So long as political or other non-hydrologically based jurisdictions are used as the fundamental planning units, important hydrologic relationships are likely to be ignored in planning, to the detriment of some or all of the inhabitants of the region. The only way to ensure that these important relationships are revealed directly and explicitly is to take a regional hydrologic viewpoint in water resources planning.

Clearly, a comprehensive hydrologic data base is needed to inform and support water resources planning in the region. This need is being partially addressed by the Water Data Banks Project of the Multilateral Working Group on Water Resources of the Middle East Process, and may be further addressed by the Joint Water Committees formed as part of the Treaty of Peace between Israel and Jordan (Appendix A) and the Israeli-Palestinian Interim Agreement on the West Bank and Gaza Strip (Appendix B). Developing such a data base will require the commitment of significant human and financial resources. Thus, for these as well as scientific reasons, it is important that the data be developed consistently and that the same measurement techniques be used throughout the study area. The need for comprehensive, consistent data for planning strengthens the case for viewing the water resources of the study area regionally.

Optimal water resources planning can only be accomplished if the study area itself is taken as the basic physical management unit. Failure to plan on a regional basis and to account for important hydrologic relationships will undoubtedly destroy important opportunities to economize on water use, take advantage of joint and conjunctive uses, manage waters of different quality best, and allocate between instream and consumptive uses optimally. In addition, as discussed in Chapter 4, cooperative water resources planning will be important to ensure that the value of ecosystem goods and services is realized.

RECOMMENDATIONS

The committee recommends that responsible national and international agencies should take a regional approach to achieve several critical ends:

1. Acquire data on water availability and uses employing consistent methodologies, techniques, and protocols.
2. Monitor both quantitative and qualitative conditions of area water resources using consistent techniques and units of measurement.
3. Encourage an open exchange of scientific research results relevant to the area's water resources and encourage relevant scientific research on a regional and collaborative basis.

REFERENCES

Al-Weshah, R. A. 1992. Jordan's water resources: Technical perspective. Water International 17(3)September:124-132.

BRL-ANTEA. 1995. Guidelines for a Master Plan for Water Management in the Jordan River Basin. Ingenierie, France: BRL. Not published.

Biswas, A. K., J. Kolars, M. Murakami, J. Waterbury, and A. Wolf. 1997. Core and Periphery: A Comprehensive Approach to Middle Eastern Water. Middle East Water Commission. Delhi: Oxford University Press.

CES Consulting Engineers and GTZ. 1996. Middle East Regional Study on Water Supply and Demand Development, Phase I, Regional Overview. Sponsored by the Government of the Federal Republic of Germany for the Multilateral Working Group on Water Resources. Eschborn, Germany: Association for Technical Cooperation (GTZ).

Cohen, S., and G. Stanhill. 1996. Contemporary climate change in the Jordan Valley. J. Applied Meteor. 35:1052-1058.

Feitelson, E., and M. Haddad. 1995. Joint Management of Shared Aquifers: Final Report. Jerusalem, Israel: The Palestine Consultancy Group and the Harry S Truman Research Institute. 36 pp.

National Research Council. 1993. Ground Water Vulnerability Assessment, Contamination Potential Under Conditions of Uncertainty. Washington, DC: National Academy Press. 204 pp.

Principal Organization of Soviet Ministers for Geology and Cartography. 1988. Geographical Belts and Zonal Types of Landshafts (in Russian) 1:15,000,000. Moscow.

Salameh, E. 1996. Water Quality Degradation in Jordan. Amman, Jordan: Friedrich Ebert Stiftung and Royal Society for the Conservation of Nature. 179 pp.

Salameh, E., and H. Bannayan. 1993. Water Resources of Jordan: Present Status and Future Potentials. Amman, Jordan: Friedrich Ebert Stiftung and Royal Society for the Conservation of Nature. 183 pp.

Stanhill, G., and C. Rapaport. 1988. Temporal and spatial variation in the volume of rain falling annually in Israel. Isr. J. Earth Sci. 37:211-221.

U.S. Central Intelligence Agency. 1993. Atlas of the Middle East. Washington, DC: U.S. Government Printing Office.

Water Authority of Jordan. 1996. The Water Authority of Jordan Annual Report. Amman, Jordan. In Arabic.

3

Factors Affecting Patterns of Water Use

The history of predicting water use and related economic activity, population growth, and other variables of importance to water and economic planners shows that precise predictions are often incorrect. Difficulties arise because there are many unknown and poorly defined variables and because people are ingenious in their adaptations to change. Although predictions, projections, and scenario building rarely provide an adequate basis for planning by themselves, they may be useful in identifying and analyzing different options. Thus, many of the factors that are likely to affect future water use can be identified.

The 1996 UN report assessing the freshwater resources of the world concludes that "water use has been growing at more than twice the rate of the population increase during this century and already a number of regions are chronically water short. About one-third of the world's population lives in countries that are experiencing moderate to high water stress, resulting in part from increasing demands fueled by population growth and human activity. By 2025, as much as two-thirds of the world population would be under stress conditions" (UN, 1996). It is recognized that areas that are short on water may in effect import it from other areas when they import food or energy or manufactured products that require water inputs.

The peoples of the study area will almost assuredly live under conditions of significant water stress during the immediate future. Barring completely unforeseen events, the population of the region is likely to grow, possibly very rapidly. Moreover, the region will likely continue to develop economically, and such economic growth could be substantial in Jordan and the West Bank and Gaza Strip. The twin phenomena of population and economic growth will place increasing pressure on the already limited water supplies of the region. The region's success in managing this intensifying water problem will largely be a matter of how it identifies and accounts for all of the many factors that determine and influence water use.

This chapter reviews the factors that are likely to influence future water use, giving special consideration to practices that may be amenable to some degree of change. The chapter's discussion begins with the difficulties of projections and the problems associated with identifying specific disparities between water supply and demand.

PROJECTED SUPPLY-DEMAND DISPARITIES AND WATER RESOURCES PLANNING

Water resource planners frequently focus on identifying potential gaps between water demand and water supply at some future date. Detailed plans are then developed to ensure that supplies are brought into balance with anticipated demands, thereby eliminating the gap. Such plans typically include projections of anticipated levels of water use based on population growth, per capita and per hectare water use, and other variables that affect demand. These estimates of future use are then compared with existing levels of available water supplies, and the time when a "gap" or disparity between anticipated levels of use and existing levels of supply is identified. Based on the size and timing of this gap, measures and actions are then identified to close the gap and bring supplies into balance with the expected demands. There are many examples of this planning methodology (California Department of Water Resources, 1994), including many of the water planning analyses for the study area (see, for example, CES Consulting Engineers and GTZ, 1996).

The committee believes that water planning approaches that initially focus on emerging gaps between supply and demand are flawed. They are based on projections that, however well informed, often ignore the considerable uncertainty surrounding future levels of water use; and more generally, such plans are always based on a significant number of assumptions, many of them untested and unstated. To the extent that water availability determines population and economic growth, projections can become self-fulfilling prophesies. By contrast, where economic and population growth are only weakly determined by the availability of water, overly optimistic growth projections may lead to investment in excess water supply capacity, the costs of which must be borne, regardless of whether the water is used. In circumstances characterized by significant levels of uncertainty, plans that are both adaptive and flexible may be desirable. The uncertain prospect of global environmental change, for example, would appear to call for flexible responses, permitting regions to adapt reasonably quickly to changes in climatic patterns as they emerge (Vaux, 1991).

Many assumptions in the traditional analyses are unstated or have not been subject to careful examination. Frequently, plans are based on the assumption that current levels and patterns of water use are optimal, irrespective of the costs of maintaining them in the face of population and economic growth. The role of prices in rationing the quantities of water used is rarely considered, since most such studies are premised on the unstated notion that real or nominal prices for water should remain constant. Additionally, planning based on the analysis of gaps frequently fails to identify the full range of adaptive mechanisms through which growth in water demands can be accommodated. Thus, such studies have often been done for the sole purpose of justifying the "need" for public financing of additional facilities, even though other less expensive means for balancing water supply and demand may have been available. It is only recently that such plans have included significant consideration of various options for managing water demand (see, for example, Berkoff, 1994).

The ultimate issue for planning, of course, is not closing any gap in the most literal sense: the quantity of water supplied *always* equals the quantity demanded or used. It is not physically possible for an individual or population to use more water than is actually available. The issue is rather that the amount of water desired or needed for many purposes may exceed available supplies. In many Middle East households, water supply is severely restricted for some or all of the time, and such restrictions may grow as economies and populations grow unless water delivery capacities can be increased. Where existing levels of water use are extremely low compared with the levels needed to avoid high economic and social costs, attention understandably focuses on closing the water supply "gap" between current and minimally acceptable levels of use. In addition, where current levels of supply include sources that cannot be sustained over the long-run—"mined" ground water, for example—it will not be possible to maintain a balance between supply and existing patterns of use indefinitely. Bringing demand and supply into balance in such cases requires reallocation among uses or the development of new supplies. Either action may entail substantial economic, social, or environmental costs.

The planning issue, then, is the identification of alternative options through which supply and demand can feasibly be balanced, the costs of achieving these various alternatives, and the amounts of water use and allocations among sectors that each alternative allows. Any plan for managing water scarcity in the Middle East should identify the full range of alternatives for augmenting water supply and managing demand, provide estimates of the costs of each alternative, and identify and characterize alternative levels of water use where supplies and demands are in equilibrium.

In developing and assessing strategies to manage scarce water supplies, it will be important to identify the variables that have a large influence on the level of water use. It will be equally important to understand the extent to which these variables can be managed and controlled, or be changed by further research and development.

FACTORS THAT AFFECT WATER USE

General Factors

The quantities of water used in any activity are jointly determined by the supply of water available to support that activity and the demand for water in that activity. Both the supply and the demand for water are further determined by variables that tend to be location specific. Nevertheless, a number of overarching factors influence levels of water use independent of location. These factors will undoubtedly be critical in determining future levels of water use in the study area.

Population Numbers and Distribution

At the most fundamental level, water is needed to supply people's basic domestic needs, in quantities directly proportional to the number of people. Other uses of water include the various municipal, industrial, agricultural, environmental, and other uses described elsewhere in this report. The quantities of water used for these purposes are also related to some degree to the number and spatial distribution of people in the region, but these quantities are also affected by many other factors, discussed below. Finally, people residing in urban areas tend to have different patterns of water use, and they tend to use different quantities of water than people in rural or agricultural areas.

Trends in population growth and distribution are extremely difficult to predict. Currently, annual population growth rates in the study area are 3.6 percent in Jordan, about 3.1 percent in the West Bank and Gaza Strip, and about 2 percent in Israel. However, the growth and distribution of population has been strongly sensitive to events both within and outside the study area. For example, armed conflicts in the Middle East resulted in three waves of immigration to Jordan and changes in policies in the former Soviet Union and its ultimate dissolution led to significant immigration to Israel. Although it appears quite likely that population in the study area will continue to grow over the next few decades, the rates and distribution of this growth are extremely difficult to predict accurately.

Technology

Technology and changes in technology may affect the availability or supply of water, demand for water and levels of water use. Industrialization, for example, typically increases the demand for water, at least initially. However, technological developments that permit users to economize on water—such developments as water-efficient indoor plumbing fixtures, closed-conduit irrigation systems like drip and microsprinkler systems, and computerized irrigation management techniques—frequently result in reductions in water use. Technical improvements that improve timing and lower costs of supply can also affect water use. For example, the construction of impoundment facilities permits control and regulation of runoff and allows more constant levels of supply. Over the last century, pumping technology improvements have made new sources of ground water available that previously could not be exploited because of their depth. On the other hand, failure to employ modern technology may mean lower quantities and higher costs of available supply.

While improvements in technology have sometimes dramatically increased the availability of water supplies, technology can also produce unwanted and unforeseen side effects. Some technology-induced or technology-influenced changes in water supply may be reversible only over time scales of thousands of years. For example, the construction of large dams (NRC, 1987, 1996), exploitation of ground water and irrigation practices (NRC, 1989) may alter water quality, regional hydrology, and water-dependent ecosystems in ways that are either impossible or prohibitively expensive to reverse on any reasonable time scale. Consequently, a complete assessment, including considerations of sustainability (and intergenerational equity), of the impacts of new and existing water supply technology should identify specifically the time domains over which the benefits and costs of the technology are likely to be borne.

Economics

Economic conditions, both within and outside the study area may affect water supply and demand. Recent declines in the world price for cotton have caused sharp declines in the potential profits from cultivation of irrigated cotton. In turn, this development has provided both the political and economic impetus to reduce cotton plantations in Israel and replace those irrigated plantations with dryland agriculture, significantly reducing water demands. World energy prices also affect the quantities of water used by boosting the price of water that must be pumped or treated before it can be used. Changes in economic conditions also affect foreign trade in many ways whose implications for water use are not always easy to foresee. Finally, economic conditions within the region will affect water supply and demand by affecting the ability of water users to pay for water, as well as the ability of producers to purchase capital and labor for activities in many industries that may directly or indirectly affect water use, including agriculture.

Environmental Conditions

Changes in environmental conditions can also significantly influence water supply and demand. Increased precipitation or decreased evapotranspiration are likely to augment water supplies and reduce the water demanded by irrigated agriculture. Increases in temperature or decreases in vegetated area or biological diversity are likely to diminish available supplies and increase the water demanded in many water using sectors. Water quality deterioration due to increased contamination levels reduces the available supply of water as surely as drought.

Changes in the environment can be directly or indirectly caused by human activities, or they can be (apparently) unrelated to human activity. For example, global climate change occurred long before humans or even living organisms inhabited the planet. Such change is likely to continue, but will be continued with global change caused by human activity. The human-induced global climate change may be pervasive and may have already occurred. Global change is likely to have significant or even profound impacts on regional water supplies and demands. However, current understanding of global climate patterns makes it very difficult to assess the impacts of such change regionally and therefore to predict how such critical variables as temperature and precipitation might change in the study area.

Discussions in this report also show how ecological conditions can affect water quality and quantity, and vice versa. Since the origins and mechanisms of these interactions are not always well understood, these changes are also hard to predict. However, the certainty that environmental change will occur suggests the need for flexible water management and allocation schemes for populations in the study area and elsewhere to respond to change as it occurs.

Instream and Withdrawal Uses of Water

In characterizing patterns of water use, one fundamental distinction is that between instream and withdrawal uses of water. The flowing or fleeting nature of water resources means that in many instances, certain uses of water do not impair its availability for further use. These uses are commonly termed instream: they do not notably alter the properties of the water nor thus the quality or quantity of water to serve subsequent uses. Examples of instream uses include most recreational uses, support of aquatic habitats and other environmental uses, navigation, and generation of hydroelectric power.

When water is withdrawn from a surface water body or from an aquifer it may be used either consumptively or non-consumptively. Consumptive uses occur when water is transformed from a state or location from which it can be used to one in which it cannot be used. Water used consumptively is not available for subsequent uses. Examples include uses such as irrigation, in which transpired water is evaporated and can not be immediately captured to serve new uses, and industrial uses in which the water is incorporated into a production. For the most part, industrial and indoor household uses are non-consumptive, however, in almost all cases the quality of the water is degraded so that some form of treatment is required before it is available for further use.

The availability of water for withdrawal use is determined, at least in part, by the proportion of the supply allocated to instream uses. While instream uses do not render water unavailable for additional use, where water supplies are scarce, instream and withdrawal uses tend to compete. The importance of water in supporting the instream uses to maintain biodiversity and environmental quality is explored in the next chapter.

Regional Determinants of Water Use

While the overarching factors just discussed explain much of the magnitude of water use, both supply and demand for water are further determined by location-specific variables. The availability of supplies depends, for example, on the costs of developing and transporting the water and of any treatment needed to ensure that the water is of suitable quality for the use. Similarly, the demand for water depends on the water-intensity of local and regional economies and—where domestic use and agricultural irrigation are important—the local climate.

There are virtually no studies of the determinants of water use in the study area. The vast majority of studies on determinants of water use in semiarid environments have been done of the western United States (see, for example, Howe and Linaweaver, 1967; Bruvold, 1988). A review of these studies suggests variables that may be important in estimating the demand for water in the study area.

Municipal and Industrial Sectors

The amount of water withdrawn and consumed by the municipal sector is in large measure a function of the population size. The need for water to supply people's basic needs for drinking, cooking, and sanitation is proportional to the number of people and their standard of living. Varying quantities of water may be used for other household purposes beyond these fundamental needs, and these quantities will also be related to the number of people, though less directly so. Changes in technology or behavior that alter levels of water withdrawn and consumed may have significant impacts on the total levels used and the proportion returned as waste. It would be useful to investigate the effects of drought on water use; reductions on outdoor uses of water and economizing on indoor uses reduced total domestic water use by 25 percent in California during the drought of 1987-1993.

Domestic household water use is also importantly influenced by the number of persons in the household. Interestingly, data have been reported that show declining per capita use rates as the number of persons living in the household increases (Howe and Linaweaver, 1967; Bruvold, 1988). Household water-using technology, such as low-flow toilets, may also be an important determinant of per capita domestic water use, as are household appliances such as clothes- and dishwashers. A working water meter and accurate metering and reporting of water use have also been shown to be important determinants of water-using behavior in the household. Outdoor use of water may account for a significant percentage of total household water use, though irrigated landscaping is far less prevalent in the study area than it is in the semiarid western United States. Finally, household income and the price of water have also been shown to be important determinants of water use (Howe and Linaweaver, 1967; Bruvold, 1988). It is generally recognized that adoption of water-saving technology and drought-resistant landscaping, programs of education, and changes in prices and pricing systems can all have significant impacts on domestic water consumption. These determinants should be considered in developing any water plans for the region.

Determinants of industrial water use and return vary from industry to industry. They are importantly influenced by the technology employed. There is clear evidence that stringent standards or regulations governing the quality of discharge waters can lead to intensified recycling of industrial water, with significant reductions in total water used as well as a reduction in the quantity of wastewater discharged. Determinants of commercial and public uses are less well studied and understood. Again, a more complete picture could be obtained by checking the effects of droughts on the use of water in these sectors. Public uses are importantly influenced by the area of open space such as parks. While the patterns of municipal water use in the study area may not be closely similar to those of the western United States, the U.S. studies are likely to provide useful information in examining determinants of municipal and industrial water use in the study area.

Agricultural Sector

The determinants of agricultural water use are generally quite well known. The greatest factor influencing evapotranspiration is solar radiation. However, different crops may have different requirements for evapotranspiration, so that consumptive water use in the agricultural sector also depends on crop type. Evapotranspiration is also influenced by other climatic variables, including temperature, humidity, and wind speed. The same crop has different evapotranspirative requirements in different climatic zones. In addition, different irrigation methods such as drip, sprinkler, and different techniques of gravity irrigation, and the employment of closed-conduit irrigation technology may also reduce the quantities of water consumed at the farm level for the same crop production. The price of irrigation water has been shown to be an important water use determinant, but the impact appears to be largely on the type of crop selected rather than the amount of water applied (Green et al., 1996). Other things being equal, high-valued crops—particularly fruit, nut, and vegetable crops—tend to be grown in areas where irrigation water prices are high. There is also evidence that high water prices result in investment in technology and management regimes that reduce water losses at the farm level.

Reductions in use at the farm level may not always translate into reductions in aggregate water use. Frequently, changes in irrigation technology or management regimes result in reductions of deep percolation and runoff. In most instances, water that deep percolates from irrigated areas constitutes ground-water recharge, and water that runs off from fields is a source of supply for someone else. As a consequence, agricultural water conservation programs need to be carefully tailored to ensure they result in true water savings and not simply savings at the farm level that reduce someone else's supply (Council for Agricultural Science and Technology, 1988).

There is less information on the determinants of both municipal and agricultural water use in the study area than is desirable. The development of water management plans and the delineation of various options and levels of balanced water supply and demand will ultimately require a better understanding of both the patterns of water use and the determinants of those patterns.

Recommendations

1. A review of studies undertaken elsewhere in the world should be carried out to identify water use patterns and the determinants of those patterns. Such a review should be very helpful in structuring similar studies for the study area.

2. The committee recommends that such studies of water-use patterns and their determinants be undertaken for the study area as soon as feasible as part of the regionwide water planning effort.

CRITERIA FOR SELECTING AMONG WATER USE MANAGEMENT OPTIONS

As the populations and economies of the study area change, different water management options will need to be developed and assessed. Levels of water use can be affected both by managing variables that affect demand and by employing new water supply technology—undertaking new development, where warranted, or developing new water supply regimes. Specific options to manage demand and augment supplies in the region are considered in Chapter 5 of this report. The particular combination of supply and demand management measures selected will significantly determine the region's patterns of water use. As we have described above, these variables make accurate, precise predictions and projections impossible. Therefore, instead of basing evaluations of plans on necessarily inaccurate projects or scenarios, the committee has identified and applied the following criteria to make a critical assessment of the various options presented in this report. In fact, these criteria are essential in assessing any water use management option, especially in a water-scarce region like the study area.

1. What is the likely magnitude of impact on available water supply? Other things being equal, options that have relatively large positive impacts on available water supply, whether individually or in aggregate, will be more desirable than those that have more modest effects. In many circumstances, programs of demand management that induce economizing on water use may have very significant impacts on the available supply of water. Options to enlarge the available supply of water are clearly not restricted to supply augmentation or development projects.

2. Is the option technically feasible? Supply augmentation and demand management options must be technically feasible. In evaluating options, care should be taken to assess whether the option is currently technically feasible, and, if not, whether it could be made so with a modest additional investment in research and development.

3. What is the environmental impact of the option? Will the option reduce or increase the quantities or qualities of other water supply sources? Does the option have adverse environmental impacts? Beneficial environmental impacts? What is the impact of the option on aquatic and terrestrial habitats? Will the option lead to losses of unique plant and animal communities or particularly valuable species? (See Chapter 4 for an extended discussion of these general issues.)

4. Is the option economically feasible and fair? What factors affect the economic feasibility of the option? Has the option proved economically feasible elsewhere? Will local circumstances have a significant impact on either the costs or the benefits? In assessing economic feasibility, all costsæincluding the costs of technical external diseconomies—must be considered. Additionally, all costs and benefits must be reported together with appropriate specific information about who will bear the costs and who will receive the benefits of any options.

5. What are the implications for intergenerational equity? For purposes of analysis, the notion of intergenerational equity can be defined in terms of three principles. The *principle of future options* requires that actions taken now do not unduly restrict the opportunities for future generations to satisfy their own needs and advance their own welfare. The *principle of conservation of quality* requires that the quality of the environment be fully maintained for future generations. The *principle of access* requires that each generation be provided with access to the legacy of the past, and that this access be conserved for future generations. Each option should be assessed in terms of how well it adheres to these three principles of equity. The concept of sustainability as expressed in intergenerational equity, is variously defined in reports by scientific and public-policy organizations. It always requires some assumption—implicit or explicit—as to time horizon, the elements of environment to be maintained, and the societal standards for the access to be conserved for a specified population. (Some representative definitions are listed in the bibliography.)

REFERENCES

Berkoff, J. 1994. A Strategy for Managing Water in the Middle East and North Africa. Washington, D.C: The World Bank. Pp. 72.
Bruvold, W. H. 1988. Municipal Water Conservation. Contribution # 197. University of California Water Resources Center, Riverside.

California Department of Water Resources. 1994. California Water Plan Update. Bulletin 160-93. Sacramento, California.

CES Consulting Engineers and GTZ. 1996. Middle East Regional Study on Water Supply and Demand Development, Phase I, Reional Overview. Sponsored by the Government of the Federal Republic of Germany for the Multilateral Working Group on Water Resources. Eschborn, Germany: Association for Technical Cooperation (GTZ).

Council for Agriculture. 1988. Effective Use of Water for Irrigated Agriculture. Task Force Report No. 113. Council for Agricultural Science and Technology, Ames, Iowa.

Green, G., D. Sunding, D. Zilberman, D. Parker, C. Trotter, and S. Collup. 1996. How does water price affect irrigation technology adoption? California Agriculture 50(2)March/April:36-40.

Howe, C. W., and F. P. Linaweaver, Jr. 1967. The impact of price on residential water demand and its relation to system design and price structure. Water Resources Research 3(1):13-32.

National Research Council (NRC). 1987. River and Dam Management: A Review of the Bureau of Reclamation's Glen Canyon Environmental Studies. Washington, D.C.: National Academy Press. 152 pp.

National Research Council (NRC). 1989. Irrigation-Induced Water Quality Problems. Washington, D.C.: National Academy Press. 157 pp.

National Research Council (NRC). 1996. River Resource Management in the Grand Canyon. Washington, D.C.: National Academy Press. 226 pp.

United Nations (UN). 1996. Comprehensive Assessment of the Freshwater Resources of the World. The United Nations, New York, NY.

Vaux, H. J., Jr. 1991. California's water resources. Pp. 69-96 in Global Climate Change and California: Potential Impacts and Responses, J. B. Knox and Ann Foley Scheuring, eds. Berkeley, CA: University of California Press.

4

Water and the Environment

The significance of the environment—including ecosystem services—to the sustainability of water supplies is often ignored in addressing the study area's water-resource planning. This chapter provides evidence, first that environmental quality depends on maintaining water quality and quantity, and second, that high-quality water supplies depend on environmental quality. To a large degree, environmental quality refers to the area's ecosystems, and without the goods and services of natural ecosystems, sustaining supplies of high-quality water for people will be extremely difficult and expensive. Environmental concerns are central to sustainable water resource planning. Water-resource planners in the study area should recognize that the relationships among ecosystem goods and services and water are dynamic and interactive.

In reviewing the relationships among these services, biodiversity, and water supply and quality, this chapter makes four major points. First, maintaining and enhancing ecosystem goods and services is essential for the economic development and welfare of the study area, especially over the medium and longer terms. This stewardship will enhance the quality of life of the study area's inhabitants; and it will maintain environmental quality, including water quality. Second, to achieve such benefits, it is essential to maintain, and where possible, restore ecosystem structure and functioning (sometimes referred to as *ecosystem integrity*). Third, biological diversity has great moral, cultural, and aesthetic importance to many societies, as reflected in laws and international agreements that express commitments to protect it. In addition, many ecologists believe that maintaining biological diversity is important to sustain ecosystem functioning, although the information on this matter is still very sparse and unclear. Fourth, all these achievements require that, in plans for providing and allocating the study area's water resources, a balance must be struck among environmental, short-term economic, and other objectives. To assess these balances and identify appropriate tradeoffs, a significant amount of new scientific information will be needed.

ECOSYSTEM SERVICES

Ecosystem services are ecosystem processes and functions beneficial to humans, primarily in contributing to the sustainability of people's lives and their intensively managed ecosystems. When activities destroy or impair the ability of natural ecosystems to provide these goods and services, the goods and services must be replaced by artificial means. Examples of such replacements are wastewater treatment plants, water filtration and purification systems, erosion control programs, and so on. Wide experience has shown that the artificial replacements for natural ecosystem goods and services are usually very expensive and often inferior to the natural ones. Because natural ecosystems provide these goods and services at no immediate financial cost, they appear to be free and their value and importance are often underestimated or overlooked entirely. For example, the value of ground water properly includes its extractive values (e.g., municipal, industrial, and agricultural uses) as well as the natural, in-situ services it provides (e.g., providing habitat and supporting biota, preventing subsistence of land, buffering against periodic water shortage, and diluting or assimilating ground-water contaminants) (NRC, 1997). To take advantage of these crucial services, they must be understood and protected.

Ecosystem services can be classified into those related to air, soil, and water. One particular service, absorbing and detoxifying pollutants, can be related to air, soil, water, or some combination of the three. Some services are global in extent and of crucial survival value, namely, the maintenance of the gaseous composition of the atmosphere, and regulation of global air temperatures and global and local climatic patterns.

Although increasing quantities of cash crops are produced on soilless substrates in the study area (using growth chambers, or "greenhouses"), soil is one universal substrate for terrestrial biological production. Soils are produced by weathering of rocks in the Earth's crust. Organisms directly affect this weathering and also mediate the effects of water and air on weathering. Thus, one important ecosystem service is production and maintenance of soil. Soil can be lost by wind and water erosion at a rate orders of magnitudes faster than it is generated. Normally, soil erosion is slowed down or even totally prevented by vegetation cover. The vegetation of the drylands, through sparse, plays a similar role, which is augmented by a biogenic and soil crust, produced by photosynthetic bacteria, algae, lichens, and mosses (Boeken and Shachak, 1994). Soil retention is linked to water-related ecosystem services, and these are directly related to sustainable water supplies.

Another important ecosystem service is the maintenance of the hydrological cycle. Plants are important for this service, which is especially valuable in drylands. Plant architecture, growth form, and phenology jointly influence the fate of raindrops (i.e., what is retained by the soil, what runs off, and what is returned to the atmosphere) and generate shade, which reduces topsoil evaporation. The overall effect of the vegetation on the water balance of the ecosystem, or even of a country or region, depends on the plant community structure. A plant community is composed of all the species populations that inhabit the ecosystem. The spatial combination of the individuals of all species in the community determines the effect of the vegetation on the water balance of the ecosystem, and effects on the water balance of adjacent and even distant ecosystems as well.

The water-related services above are "input" services, which include soil moisture recharge and retention, aquifer recharge, and control of soil salinization and erosion. With respect to "output," one important ecosystem process is returning water to the atmosphere. On a global dimension, this process is clearly a service. However, on the local and regional scale in dryland countries like those of the study area, this is more a "disservice." The balance of this "service"/"disservice" is not known. For Israel, Stanhill (1993) calculated that 10,000 years ago, when the dry sub-humid part of the country (receiving 400 to 800 mm annual rainfall) was mostly a natural, scrubland ecosystem, the potential water yield (volume of rain falling in a given year on a given surface area, minus volume of water returned to the atmosphere from the same area and year) was 1,590 km^3/year, lower than the current 1,846 km^3/year, with most of the area consisting of cropland, a highly managed ecosystem. Natural scrubland ecosystems appear to evaporate more soil water than intensively managed ecosystems in Israel. However, the positive contributions of that scrubland and other less managed ecosystems—such as scrubland's contribution in recharging aquifers—to the water balance of Israel, compared to the contributions of intensively managed ecosystems, must be calculated too, and weighed against the losses due to evapotranspiration from the same ecosystems.

Services Provided by Water Bodies

Open-water body ecosystems are spatially more homogeneous and better delimited than most terrestrial ecosystems. Being mostly a dry-land, the study area is inherently poor in water bodies. Furthermore, many of these aquatic ecosystems are under intensive management or have been totally replaced by terrestrial ones. The following section addresses ecosystem services of the study area's streams, lakes, and wetlands.

Streams

Currently, the most significant ecosystem service of streams is the natural treatment of wastewater. The wastewater-treating service of most of the aquatic organisms in streams is facilitated by the oxidizing properties of the stream current and its velocity. Other components of the food web, such as aquatic herbivores and predators, are instrumental in regulating the populations of these wastewater-treating species, and in this way become involved in the quality of the wastewater treatment service of streams.

Lakes

Of the two study area's major lakes, one (the Dead Sea) is globally unique in its apparent lifelessness, and the other (Lake Kinneret, Lake Tiberias, or Sea of Galilee) serves as an operational open water reservoir for supplying water of drinking quality to most parts of Israel, with recent allocations to Jordan and the West Bank and Gaza Strip. The "service" of this ecosystem is thus to store water and to help maintain its quality as drinking water. Lakes in general, including Lake Kinneret/Lake Tiberias/ Sea of Galilee, also provide the ecosystem service of wastewater treatment, although not as effectively as streams.

Wetlands

Wetlands are lands where the water table is usually at or near the surface, or lands covered by shallow water, that have characteristic physical, chemical, and biological features reflecting recurrent or sustained inundation or saturation (Cowardin et al., 1979; NRC, 1995a). Most major wetlands of the study area have been drained totally (coastal wetlands of Israel) or partially (Hula in Israel, Azraq in Jordan). Others, especially around the Dead Sea, are still relatively intact, though small. Wetlands are characterized by the slow rate of water movement in them. This feature reduces their oxidizing capacity, making them ineffective in waste-water treatment. However, the slow water movement promotes the deposition of suspended material and provides ample time for the complete biological mineralization of organic compounds and biodegradation of synthetic toxic chemicals (NRC, 1992). The slow water movement also supports typical wetland vegetation, which further slows water movement, and reduces the depth of the wetland, thus contributing to its spatial expansion. This expansion provides a unique ecosystem service: water storage during floods and a slow downstream release. Wetlands therefore lower flood peaks and their detrimental economic and environmental effect, such as soil erosion (NRC, 1992). While this service is not provided by landlocked wetlands such as Azraq, it was an important (although underestimated) function of the Hula wetland before its drainage.

Artificial Aquatic Ecosystems

All types of artificial open water bodies function as intensively managed ecosystems. These bodies include fish ponds (mainly in Israel), wastewater treatment plants (e.g., the Shifdan plant in Israel), water carrying systems' open canals and reservoirs (the National Water Carrier in Israel and the Ghor Canal in Jordan), and other open air reservoirs (e.g., floodwater reservoirs in Jordan and Israel). Soon after construction, such bodies are colonized by aquatic microorganisms, plants, and invertebrates, and they are used by waterfowl and insectivorous bats (Carmel and Safriel, 1998). Thus, the water bodies, constructed for the sole function of water treatment or supply, become intensively managed ecosystems, with ecosystem functions shaped by the wild species that successfully colonize them. Like natural and less intensively managed ecosystems, constructed aquatic ecosystems provide the ecosystem service of promoting wastewater treatment. Many of these water bodies are also important habitat for birds, especially birds that migrate or that use it for wintering. Constructed wetlands for wastewater treatment can also provide wildlife habitat (U.S. EPA, 1993). In Israel, the major wastewater treatment facility, Shifdan, has become a waterfowl sanctuary that attracts hundreds of bird-watchers every year and is used to university teaching. Nearly all constructed water bodies in the sudy area significantly support bird and other aquatic and riparian biodiversity.

BIOLOGICAL DIVERSITY

Biological diversity means the diversity of genotypes within a species, species diversity, and the diversity of ecological communities: in short, biological diversity (often abbreviated to *biodiversity*) is the diversity of life on Earth. (See the similar definition by the 1992 United Nations Convention on Biological Diversity [Anonymous, 1992]). The protection of endangered species and biodiversity in general has been important to many people for a long time, and as a result, many have looked to science to provide quantitative assessments of the value of biodiversity. Although that endeavor has been difficult, there are other good reasons to protect biodiversity. For example, Sagoff (1996) described how difficult it is to establish on a purely economic basis that biodiversity or indeed most individual species should be protected, but he argued that the best reasons—and they are very powerful—for protecting biodiversity in most cases are ethical, moral, cultural, and aesthetic. Societies around the world have, in their laws and international treaties, reflected this view. Thus the United States's Endangered Species Act of 1973 declares it "to be the policy of Congress that all Federal departments and agencies shall seek to conserve endangered species and threatened species..." (Section 2 {b [c]}). The act further specifies that the determination as to whether a species is endangered or threatened must be based "solely on the basis of the best scientific and commercial data available," i.e., without reference to economic considerations (Section 4 {b[1]}) (see NRC 1995b for a description and history of the act and its scientific underpinnings). In the study area, Annex IV of the Israel-Jordan Peace Treaty (see Appendix A) includes commitments to protect natural resources and biodiversity; the presence of parks and other protected areas throughout the study area is further indication of the study area's commitment to protecting biodiversity.

The above discussion does not suggest that biodiversity has no economic value or that it is not important in maintaining ecosystem goods and services. Clearly, some *species* have enormous economic value and ecological importance, and some ecosystems have economic—especially tourism—value because of their biodiversity. Within the study area, several ecosystems have recreational and hence economic value. For example, woodland ecosystems are relatively rare in the study area, but their sharp contrast with the more common deserts make them important recreationally and inspirationally. Aquatic ecosystems are even more valuable in these respects, especially when they occur in deserts, such as the Azraq Oasis or wetlands and oases around the Dead Sea. Lake Kinneret/Lake Tiberias/Sea of Galilee, although it is in a relatively fertile area, is a major site for tourism and leisure activites, especially in summer. On the other hand, the study area's deserts and their own biodiversity contrast sharply with the landscapes that are home to most foreign tourists in the study area, and hence deserts are major sources of tourist income.

Without some minimum amount of biodiversity, ecosystems would function poorly, even if the general relationship between biodiversity and ecosystem functioning is unclear (see Grime, 1997 for a clear summary of this matter and citations to recent literature). In addition, many have cautioned that just because we cannot at present quantify relationships between biodiversity and ecosystem functioning, that does not mean we should be cavalier about extinctions: a species lost is gone forever, and we might discover too late that it had great ecological or economic importance (e.g., Perrings, 1991). Sagoff (1988) warned that if we wait to establish the economic and survival value of biodiversity, it may be irreversibly lost.

For all the above reasons, the committee concludes that it is important to protect biodiversity, and that water-resource planning should take this into account. Furthermore, protecting biodiversity often requires the protection of ecosystems, as does the protection of ecosystem goods and services. Thus, maintaining biodiversity and ecosystem goods and services can often be treated as a single goal, as the committee does in the following sections of this chapter.

Economic Values of Individual Species

In general, economic values of species derive from their provision of food, fuel, and fiber. In addition, some species provide medicinal, ornamental, and aesthetic goods. All human food consists of species and their direct products. Most of the species consumed by the global human population are domesticated and cultivated, which is also to say derived from species provided by biodiversity, or wild species. Many domestic species, and especially the food species, do not exist anymore in non-manipulated ("natural") ecosystems, namely, in the wild. But their progenitors, and more often their wild relatives, still occur in natural ecosystems. It is the genetic diversity of these progenitors and relatives that is one of the most critical benefits of biodiversity.

Ironically, domestic species are the most endangered species, despite the huge sizes of their populations and their large geographical extent. Efforts to increase their production have led to erosion of their genetic variability. These species gradually lose their resistance to environmental changes, competitors, pests, and parasites. The high densities and uninterrupted spatial expanses of their populations, and the "globalization" that leads to widespread uniformity of their genetic structure and to high transmissibility of their mortality agents, make them increasingly prone to extinction (Hoyt, 1992). Their wild progenitors and relatives provide a repository of transferable genetic variability, variability that can counteract the ongoing genetic erosion of the domestic species, thus reducing their extinction risks.

The study area of this report is one of the Earth's richest areas in progenitors and relatives of domesticated species (Zohary, 1983). Land uses depending on supplies of irrigation water deny these biogenetic resources their natural habitats where they dynamically evolve under our changing environment (Zohary, 1991). Because this genetic diversity is the insurance against agricultural disasters, its loss through excess water use jeopardizes the long-term sustainability and contributes to non-sustainability in the use of the study area's water supplies.

Many wild species, both terrestrial and aquatic, are of commercial value. Wild plant species are often heavily sought, collected whole or for their parts, for their herbal, aromatic, medicinal, and ornamental properties. Many wild plant species are labeled prime pasture species, because they are critical for range-dependent livestock. Expansion of irrigated agriculture is at the expense of this economically significant biodiversity.

At the same time, most species do not have current, short-term economic value. Of the approximately 266,000 species of plants known (Raven and Johnson, 1992), about 5,000 are used as food plants, 2,300 are domesticated, and 20 provide most of the food for the global human population (Frankel and Soulé, 1981). Food production is currently limited by land and water resources and losses to pests, but not by the lack of food species. However, should current food species fail because of the risks identified above, alternative species, currently wild, will be sought for domestication. The natural species pool is thus a repository of potential food and utility species for humans.

Farmers often view the natural ecosystems adjacent to their croplands as sources of pests. But these and other natural ecosystems are also, and sometimes mostly, sources of enemies of agricultural pests. Thus, natural ecosystems provide important services that have economic value. Use of synthetic chemicals to control pests also controls their enemies, so this potential ecosystem value (pest control) is often not realized.

Conflicts Between Water Resource Development and Ecosystem Goods and Services

All species of realized or potential economic benefit to humans, globally and in the study area, are land users, and this type of land use competes with irrigated cropland. Improved water supplies for the study area may reduce not only the economic benefit of any expanded agriculture, but also the sustainability of existing irrigated and rain-fed agricultural production. This potential conflict requires evaluating that part of biodiversity that is of economic significance, but even the fraction and magnitude of that part of biodiversity have not yet been anywhere identified (Lawton, 1991). All species and their different populations must be considered possible members of this economically important class, at least until a large part of the useful species are identified as such. Their benefits must then be weighed against the benefits of developments driven by water supplies in the study area.

Water Supplies, Biodiversity, and Desertification

There is a critical relationship between ecosystems, desertification, loss of biodiversity, and climate change in the context of sustainable water supplies. Desertification is land degradation in drylands caused by mismanagement and overexploitation. Overpopulation and increased demands, mostly in semiarid regions, bring about overstocking and overgrazing and trampling, transformation of woodland to rangeland (e.g., the deforestation for railway ties and fuel in the study area by Turkish forces in the early part of the century) and the overexploitation of rangeland for the fire wood. The reduced vegetation cover and breakage of the soil crust, lead to water and wind erosion of the topsoil, and with it an irreversible loss of productivity-desertification. The loss of vegetation cover reduces aquifer recharge and increases losses of floodwater. At the same time the loss of vegetation cover reduces the global carbon sink, thus exacerbating global warming.

Another type of land degradation is associated with the transformation of rangelands with year-round vegetation cover, to croplands that if not irrigated, have only intermittent cover, leading to further soil erosion. If the croplands are irrigated, irrigation brings about salinization of the topsoil: water scarcity does not permit application of quantities sufficient for leaching, and the high evaporation leaves the salt in the topsoil. Such croplands, when abandoned due to salinization, cannot revert to their original function as rangeland, since most range species are intolerant of the increased salinity. Thus, either due to loss of topsoil or due to salinization or both, land degradation may reach the point of irreversible desertification. To conclude, increasing the water supply allows the intensified use of rangelands and their conversion to croplands in semiarid regions. This leads to loss of biodiversity, reduced ecosystem services such as soil conservation, aquifer recharge, and the maintenance of carbon sink, thus exacerbating desertification and global warming.

Desertification often has roots, typically a large external disturbance (Puigdefabregas, 1995), that began some years, or even decades, before crises manifest themselves (e.g., the Dust Bowl in the United States in the 1930s and the Sahel crisis in the 1970s). For this reason, it is important to try to prevent desertification by avoiding non-sustainable use of water, before it manifests itself by loss of biodiversity and the impaired provision of ecosystem services such as aquifer recharge, leading to reduced sustainability of water supplies.

ENVIRONMENTAL COSTS OF
WATER-RESOURCE DEVELOPMENT

Policy makers, planners, and individuals in the study area need to make many decisions about activities ranging from international development projects to individual actions concerning water use, waste disposal, and what to plant in a garden or field. To make decisions about these activities and allocate water resources to different uses in the study area, a balance must be struck among environmental, economic, and other objectives when those objectives do not represent the same uses of water. To assess the balance and to identify acceptable tradeoffs, current scientific information should be used; a significant amount of new scientific information will also be needed.

The preceding sections explained how environmental quality depends on the goods and services provided at no cost by natural ecosystems and explained how economic well-being, quality of life, and maintenance of water supplies depend on environmental quality. This section describes some of the specific consequences that follow from failure to maintain ecosystem goods and services by losing the land that is needed for ecosystems to persist.

The section illustrates some of the factors that must be considered in making assessments and identifying sound tradeoffs, by describing interactions among environments, ecosystem goods and services, water quality and quantity, and human activities in the study area. First, we characterize the study area's biodiversity in the context of water supplies, we then describe the effects of water-resource development in biodiversity and on ecosystem services; and finally, we address ways to mitigate negative effects in achieving sustainable water supplies for the study area.

Biodiversity of the Study Area

Biodiversity relevant to water use in the study area has the following features:

1. The ecology of the study area as a whole is that of hyperarid, arid, semiarid, and dry subhumid dryland ecosystems. The area's biodiversity is therefore that of drylands, with terrestrial vegetation directly limited by water, and all other components of biodiversity affected directly or indirectly by the variability and the unpredictability of water availability (Noy-Meir, 1973).

2. Humans have had an extremely long and persistent influence on the environment. It is highly likely that the recent evolution of the study area's biota took place in the presence of humankind, and that human activities and practices have acted as selection agents, like other agents of natural selection

Attributes (1) and (2) imply that many of the study area's species have been selected to withstand water scarcity, fluctuations in water supply, and human interventions, and hence that the study area's ecosystems are resistant and resilient.

3. The study area is not only a crossroads of continents (Africa, Asia, and Europe), but also of biogeographical regions—the Saharo-Arabian (African), Irano-Turanian (Asian), and Mediterranean. The area also shows intrusions and relicts of Euro-Siberian (northern European and Ethiopian (tropical African) species.

Attributes (1) and (3) have three implications: overall species richness is very high; most species are presented by peripheral populations, and, although most of the species are not unique (endemic) to the region, the communities, that is, the regional assemblages of species, are. In this region, species of Asian steppes interact in the study area with species of Saharan deserts, for example.

Effects of Water Use on Regional Biodiversity and Ecosystem Services

It is clear from the preceding that biota and ecosystem services depend on water. Water-resource development in the study area usually entails six major practices: transportation of water from lakes and river sources; pumping from sources of springs as well as impounding springs by enclosing them in concrete structures; drainage of wetlands and large ponds; drainage of ephemeral ponds; pumping from aquifers; and damming floodwater courses to construct floodwater reservoirs. Each of these practices has notable effects on biodiversity and ecosystem services of the area, as discussed below.

Management of Lakes and River Sources. Large water-development projects have dramatically affected the regional economy by promoting year-round intensive, pressure-irrigated agriculture as well as urban development. These development projects are associated with the management of river systems. The coastal Yarkon River[1] fed by Ein Afek springs at the Judean foothills generated the Yarkon-Negev Line. The Rift Valley's Jordan River Basin management generated the Israeli National Water

Carrier System and the King Abdullah Canal. These projects revolution-
ized agricultural and rural development at the cost of biodiversity and
ecosystem services, many of them related to water quality.

As Appendix C shows in greater detail, a watershed-management
approach for the whole Jordan River Basin could be essential to achieving
overall sustainability of water-resource development in the area. This
approach could achieve a sound balance between providing water supplies to the study area, and maintaining and promoting ecosystem services related to water quality—those of both the Hula wetlands and Lake
Kinneret/Lake Tiberias/Sea of Galilee—as well as the biodiversity of the
lower Jordan River and around the Dead Sea coasts.

Impoundment of Springs. Springs in the study area vary from those
sustaining small ephemeral or even permanent ponds, to those supporting ephemeral or permanent streams. Many springs are pumped and
impounded within a sealed concrete construction, to prevent evaporation
and to protect the pumps from vandalism. These practices affect both the
riparian biodiversity along the stream, mostly of plants, and the aquatic
biodiversity of the ponds and streams themselves, mostly of invertebrates,
some of which are unique to the study area. The effect of drying streams
and obstructing access to ponds also cascades to the terrestrial biodiversity
adjacent to the springs and streams, and ultimately even farther.

Drainage of Wetlands and Ponds. At the end of the nineteenth century
there were 200,000 dunams of wetlands[2], west of the River Jordan, 97
percent of which have now been drained. The motivations for this drainage were to reduce water loss by evaporation, collect water for agricultural development, increase land resources for agriculture, and eradicate
malaria. Management intensity has ranged from complete transformation of the wetland to agricultural land, with a total loss of aquatic
biodiversity, to initial draining, with subsequent engineering of a wetland
nature reserve in part of the drained area and final reconstruction of
another part by reflooding. The effects of draining wetlands, too, cascades to adjacent and remote ecosystems, by affecting animals of an amphibian life style, animals who prey on aquatic organisms, and so forth.
The Kebara wetlands on Israel's coastal plain were used by the prey of
prehistoric humans (Tchernov, 1994) and inhabited by the Nile crocodile
at the turn of this century. Also, the study area serves as a flyway of

---

[1]This river has been called "Yarkon River" for the last 60 years on international maps.
However, it is also known as Wadi Abu Butrus.

[2]A dunam, 1000 m^2, is a unit of land used in the study area.

Palearctic migratory birds, used just before or after crossing the Sahara desert. The study area's wetlands, either at the edge of deserts (the Hula) or well within them (Azraq) function for refueling prior to the desert crossing (fall migration), or rights after completing it (spring migration) (see Boxes 4.1 and 4.2). Thus, the loss of these wetlands affects European and African birds, and may modify the birds' patterns of cross-desert migration.

Loss of Ephemeral Ponds. The study area's rainy winters generate ponds that often completely dry out during the long, dry summers. Many were created by ancient damming and quarrying, and were used for generations to water livestock in early summer. They harbor unique biodiversity, adapted to the ephemeral conditions, usually by having an amphibian lifestyle or leaving dormant propagules in the soil of the dried-up bottoms of the ponds. When wet, the ponds attract wildlife that come to drink or to prey on other animals. Most of these ponds have been cut off from their runoff sources and drained, to be transformed to agricultural land. Other ponds have become sinks for wastewater of high toxicity or high organic load. Many have been drained intentionally or are sprayed to control mosquitoes. Spraying of existing ephemeral ponds, and their spatial rarity, which prevents migration between them, have reduced their biodiversity. For example, Hadera Pond had 56 aquatic plant species in 1906, of which 30 persisted until the 1950s, and only 16 in 1982 (Mador-Haim, 1987). Implicating the ponds as a mosquito threat is flawed, because mosquitoes are controlled by the ponds' natural predators—tadpoles in the winter and predatory insects that live in the ponds as long as they have water. These animals maintain mosquito populations at low levels. The use of pesticides to control mosquitoes aggravates the situation: their natural enemies are destroyed, and the mosquitoes evolve resistance to the pesticides. Fortunately, because of the dormancy and high dispersi-bility of the propagules (airborne or transported by birds), the biodiversity of such ponds can be established and promoted once they are reconstructed, by seeding the constructed ponds with soil from the few existing, healthy ponds. The most important ecosystem services of these pools are recreational, educational, and scientific, given the unique nature of their biodiversity and their dynamic ecology. The interest in Israel's ephemeral ponds has generated surveys and plans for rehabilitating and constructing ephemeral ponds, rather than permanent artificially maintained water bodies (Gazith and Sidis, 1981).

Pumping from Aquifers. There is no evidence that lowering the water table through pumping aquifers in the study area poses risks to terrestrial biodiversity. Such pumping was blamed for the deaths of acacia trees in the Israeli section of the Arava Rift Valley (Ashkenazi, 1995), but the issue is not yet resolved (Ward and Rohner, 1997). Pumping from aquifers, however, can reduce the discharge of springs, and thus transform permanent spring pools into ephemeral ones or curtail the flow of streams. For example, pumping from the Yarkon-Taninim aquifer reduced the discharge of the Ein Timsah spring into the Taninim River[3] such that the flow in the lower part of the river was reduced from about 84 to 88 million m^3/yr between 1953 and 1956 to 24 million m^3/yr between 1984 and 1986 (Ben-David, 1987). This and other factors have adversely affected the biota of this stream, despite its legal status as a nature reserve.

Damming Runoff Courses and Constructing Reservoirs. The objective of dams in the study area is to prevent runoff to either the Mediterranean Sea or the Rift Valley. The floodwater is stored in reservoirs or used to recharge aquifers or directly for irrigation. Unlike all other practices, which have a strong local effect (mostly on aquatic and riparian biodiversity), and a smaller regional effect on nonaquatic species, this management practice has a regional, whole-watershed effect, mostly on terrestrial biodiversity. The closer the dam is to the water divide, the larger the area of watershed affected. Dams and reservoirs can promote very successful agriculture, but also adversely affect downstream channels in various ways. Dams can reduce soil moisture in these channels and their immediate surroundings, adversely affecting the richest parts of dryland watersheds. In hyperarid dryland watersheds, the channel is the only element that has perennial vegetation. Putting dams in these drylands has a stronger effect on biodiversity than putting them in less arid drylands of the study area. Damming also reduces the subsurface runoff in the channel, which lasts longer than the surface runoff and is critical for the persistence of the channel vegetation. Finally, the reservoirs enrich the desert with water bodies that dramatically affect the behavior, population dynamics, and structure of the desert's annual-plant communities. These changes may be exacerbated since the development of aquatic biota in the reservoirs often leads to introducing predatory fish to control mosquitoes. While the fish are ephemeral, they attract birds and thus generate instabilities in the bird populations and in their roles in ecosystem functioning.

Other effects of runoff on wadi beds and their surroundings depend on those features of the flow that vary spatially and temporally, from flash floods causing severe erosion and loss of organisms to moderate currents that leach salts and deposit nutrient-rich soil. Dams change the

---

[3] This river has been known as the Taninim River for approximately 60 years on international maps. However, it is also known as Wadi Zarka.

BOX 4.1 The Azraq Oasis (Jordan)

The Azraq oasis or wetland is approximately 80 km east-southeast of Amman in the heart of the Azraq basin, which covers about 12,710 km^2 and is located in southern Syria, eastern and central Jordan, and northern Saudi Arabia. The Azraq oasis is an outstanding example of an oasis wetland in an arid region. The oasis is fed by springs known locally as the Shishan and Drouz springs. The oasis is famous for its date palms and provides important habitat for migratory bird species as well as mammals such as the oryx.

Water was withdrawn from the oasis to serve the needs of Amman's growing population beginning about 1960, and the rate of withdrawal increased until about 1990 (Ramsar Convention Bureau, 1994). The current condition of the Azraq oasis is typical of arid region wetlands throughout the Middle East. Extensive pumping of the upper aquifer complex to provide water supplies for municipal, industrial, and agricultural uses has caused the springs that feed the wetland to dry up.

The characteristic land form within the basin is a rolling gentle plateau with minor relief in and around the central part of the basin. Elevations range from 1,576 m near Jebel El-Arab to 500 m at the Azraq depression. Many intermittently flowing wadis drain from all directions into the Azraq depression, forming ephemeral ponds that may remain for several months before being lost to evaporation. The basin overlies three aquifer complexes—upper, middle, and lower—with the waters in the upper aquifer complex being the most utilized. The basin is a major source of drinking water for Amman, Zarqa, and Irbid, three of Jordan's major cities.

The climate of the Azraq basin is characterized by hot dry summers and cool winters, during which most of the modest precipitation occurs. Mean annual rainfall ranges from 300 mm in the north to 150 mm in the west to less than 50 mm in the south and east. Precipitation comes from cyclones that arise in the west, cross the Mediterranean Sea, and bring cool air masses from Europe. The cyclones frequently spawn thunderstorms, producing rainfall that is irregular in intensity and duration. Convective or thunderstorm rainfall is oriented along the major axis of the eastern Mediterranean and extends eastward into Iraq. This type of rainfall contributes most of the precipitation in the Azraq Basin and occurs mainly during October, November, April, and May. More widespread, gentler rains (cyclonic rains) also reach the area, mainly in December, January, February, and March.

In 1987, at a meeting of the Conference of Contracting Parties in Regina, Canada, the parties concluded that the current rate of water withdrawal, 16 million m^3/yr, was "likely to provoke serious changes in the natural properties of the Azraq wetland and in particular [could] increase the salinity of the remaining water there." The conference called for a "proper assessment of the environmental impact of the pumping" and suggested that pumping be reduced by at least 50 percent, as recommended by Jordanian conservation organizations, at least until the environmental assessment was complete. It also urged that there be a long-term plan for water use that guaranteed the natural properties of this internationally important wetland (Ramsar Convention Bureau, 1994).

However, by 1990, the Jordanian Government established a "safe yield" of greater than the 1987 rate of pumping; this was 20 million m^3/yr. The Jordanian Government invited the Ramsar Convention Bureau to apply its monitoring procedure at the site. This was done in March of 1990, and led to the conclusion that to restore the ecological character of the oasis, water withdrawals needed to be "reduced considerably." The convention made several other recommendations, including research recommendations.

After this procedure, the Jordanian Department of the Environment hired consultants to undertake studies of groundwater management, and the Global Environmental Fund (GEF) of the United Nations Development Program allocated $US3.3 million for five subprojects over a three-year period. They are intended to halt degradation of the Azraq wetlands and establish a management plan to allow the water resources of the Azraq basin to be used on a sustainable basis while preserving the unique biodiversity that characterizes the Azraq wetlands.

Although the project is in its early stages, an effort will be made to eliminate the overdraft from local aquifers and manage those aquifers so that water can be extracted from them to maintain the wetlands on a sustainable basis. A monitoring and assessment program has been included in the plans for restoration, to ensure that the restoration process achieves the goal of preserving the wetlands on a long-term basis. Should this restoration plan prove unsuccessful due to the difficulties of bringing ground-water extractions to an equilibrium (to make water available to support the wetlands), consideration will need to be given to developing imported supplies to support the wetlands. Imported supplies will be considerably more expensive than local ground-water supplies. Unfortunately, the ecological condition of the oasis continued to deteriorate for a while after 1990, and it is too early to determine the results of the GEF investments (Ramsar Convention Bureau, 1994).

Experience with the Azraq oasis shows how overexploitation of water resources in arid or water-scarce regions can threaten or destroy unique ecosystems. This experience has been repeated in many areas of the world, and as described here, in the Hula Wetlands of Israel. These ecosystems provide important services and habitat. Such unfortunate results are more likely when wetlands are treated as public goods with no one having an incentive to ensure that the services of the wetland are maintained and preserved. The case also illustrates that proactive investment and management strategies are required if unique environments such as the Azraq wetlands are to be protected and restored. Such strategies must provide for continuing management and monitoring to ensure that the wetlands resource is preserved and sustained.

BOX 4.2 The Drainage of the Hula Valley (Israel)

Background

The Hula Valley is located in the north of the Jordan Valley rift valley, in the watershed basin of the Sea of Galilee (Lake Kinneret or Lake Tiberias). The valley outlet toward the lake was blocked by a volcanic eruption many years ago. Since then drainage out of the valley is limited, and conditions favoring flooding have prevailed. In the 1950s the center of the valley was covered by a lake (of about 10 km^2), and a significant portion of the area, mostly north of the lake, was covered by marshes. An area totaling about 70 km^2 was covered by water year-round and could not be used for agriculture. The marshes were breeding grounds of mosquitoes and the whole region was infested by malaria.

The lake and surrounding wetland were a unique ecosystem. The area was covered by a rare combination of plants representing cold and warm regions and was inhabited by numerous native animals as well as birds staying until their yearly migration. The lake and marshes were drained to help eradicate malaria and to expose the land to agricultural development.

The War Against the Swamp

Social Attitude

Israeli society in the 1940s and 1950s was a society of "pioneers," who were motivated by the desire to overcome the harsh natural obstacles of the country. Like most other Western civilizations, increased productivity has been a central societal value, and "emptiness" or undeveloped land was considered undesirable. A term describing this "emptiness," *yeshimon* in Hebrew, implies a destroyed, devastated environment. The positive activity of the pioneers was reported as "the struggle of man with the emptiness."

The Hula swamp represented the evils of nature. It was hardly penetrable, and disease-carrying mosquitoes inhabited it. Thus, the desire and ambition to win the battle against the swamp and to turn it into a site where people could live and prosper was deeply embedded in the Israeli social attitude at that time. Very few people had sentiments or consciousness toward the Hula as a natural resource. Nature and natural attractions were not considered at the time as an economic asset. Tourism hardly existed and the meager internal tourist business was centered on resorts where people came to rest from their work.

A few experts from Israel and abroad expressed some hesitation about the possible hydro-geochemical processes that might follow drainage of the Hula, such as land subsidence. These warnings were not taken seriously. A fraction of the area was not developed and was later reconstructed to function as a wetland nature reserve. This area was the least suitable for a natural reserve, at an elevation above the rest of the area, making it difficult to keep the land flooded. It was also close to a major drainage canal and therefore became polluted.

Environmental Changes (1955-1970)

Different types of organic soils, from muck to peat, have developed in the marsh area. Soil organic matter is highly unstable under the prevailing climatic conditions; it could be stable only under water. Once it was exposed, very quick decomposition processes took place. These processes, together with mechanical change in the structure of the organic colloids induced by desiccation, led to the subsidence of the organic soils. The area covered by organic soils subsided at an annual rate of 7 to 10 cm. This changed the drainage system continually, shifting the elevations of different zones in the valley, requiring a continual deepening of drainage canals.

The drained organic soils caught fire from external sources of ignition and through spontaneous combustion. When ground-water level is lowered, exothermal decomposition processes take place below the thermal insulation of the dry surface, leading to extensive heating and finally to combustion. The fires created immediate damage, causing air pollution and destroying farm land. Decomposition of the organic matter led to the release of several by-products, such as ammonium, and the subsequent accumulation of nitrates in the soil. Winter flooding of the area led to the washout of nitrates through the Jordan River to Lake Kinneret. The reclaimed marsh area was the major source of nitrates into the lake, apparently starting a sequence of processes leading to eutrophication of the lake. The nature reserve did not preserve the flora and fauna of the lake and the marshes. Many species disappeared, and an area once covered with a variety of plants and animals became a flat dusty region.

The Hula Valley Management Reconsidered: 1970-1977

Organizational Aspects

The collapse of the Hula hydrochemical system, as reflected in the increase in the flux of nitrates to Lake Kinneret/Lake Tiberias/Sea of Galilee and the early indications of eutrophication, led to a strong public response. The government decided to establish a watershed authority, the Lake Kinneret Authority, to coordinate all the efforts related to the management of the lake. Another coordinating body, the Hula Committee, was established to coordinate the work in the Hula Valley specifically. These organizations were the first environmental authorities in Israel and paved the way toward the development of other environmental authorities and finally to the establishment of the Ministry of the Environment.

A unique feature of the Lake Kinneret Authority is that it was made up of representatives of municipalities, farmers, water authorities, and industries in the watershed. The traditional conflict between business and production, on one hand, with environmental authorities on the other, was thus avoided.

The government regarded the maintenance of high water quality in the lake as a priority goal and allocated a relatively large budget for this purpose. The first act of the Hula Committee and the lake authority was to support and coordinate research aimed at the solution of the nitrates problem in the Hula Valley. A number of academic institutions were represented in a multidisciplinary research team. Frequent contacts and real-time feedback were maintained among scientists, managers, and engineers of the lake authority. As a result, the research was practical mission-oriented work, with many mid-course corrections and adaptation, and the implementation of the research findings was immediate.

BOX 4.2 (continued)

Major Research Findings

A detailed study of the basin's hydrology was conducted. An artesian hydraulic gradient was found to induce a slow upward seepage of ground water, preventing any seepage of nitrate-rich water downward. Nitrate leaching occurred during the rainy season and during flash floods common in the Jordan River. During floods, when the water level in the river rises, water would flow into the low-lying peat area. A backflow of water following the flood carried large amounts of nitrate to the river and into the lake.

The microbial processes leading to the production and destruction of nitrates in the basin were studied. The effects of temperature, moisture contents, and oxygen on the production of nitrates was defined and quantified. The rate of nitrate accumulation was found to be high during the summer, if a deep soil profile was aerated and drained. The conditions to maximize denitrification, a process that lead to the conversion of nitrates to inert N_2 gas were studied in detail.

This process occurred at a very fast rate when a dry soil was rewetted quickly. A burst of microbial activity was induced, due to the exposure of fresh surfaces of organic macromolecules disengaged when hydrogen bonding was reduced in the desiccated soil. It was found that properly timed sprinkle irrigation was an effective way to reduce the nitrate content of the soil. Intensively growing forage crops were found to take up high amounts of nitrogen and slowed nitrate accumulation.

Implementation

The research findings indicated that proper water management was the way to minimize nitrate production and leaching out. By deepening and widening the riverbed and constructing dams and banks, water entry during floods was minimized. Ground-water levels during the summer were maintained at a depth of 60 to 80 cm. Water level was reduced toward the winter, to about 120 cm, to make place for the rainwater. Water level was maintained using a series of dams on the main drainage canals, connected by a network of smaller canals.

The previously common subsurface irrigation system was replaced by a sprinkler system. The properly spaced and controlled irrigation-induced intensive denitrification reduced the nitrate storage in the soil profile by an order of magnitude. This operation can be considered the first field-scale biotechnological project. The above-mentioned steps were implemented within five years following the start of the work. The project led to a marked reduction in the leakage of nitrates from the region and in the influx of nitrates and total nitrogen into Lake Kinneret/Lake Tiberias/Sea of Galilee.

Environmental Changes (1977-1990)

The intensive environmental research conducted during the previous phase produced many by-products. One was a series of results that helped to improve the agricultural production in the valley, through improved irrigation methods, improved fertilization, soil cultivation, and species selection. This success encouraged the agricultural use of the area. The land that belonged to a governmental holding company was divided into individual cooperative settlements. Another factor that affected the agricultural use of the area was the relatively high price of cotton and the profitability of cotton farming.

An additional change was the slow transfer of control of the area and the water system from the lake authority to the local water authority, dominated by the agricultural sector. The experts that acquired knowledge and understanding of the system were not involved in the control of the area. The previously recommended control of high ground-water level and the dense drainage-canal system needed for such control posed a difficulty for the farmers. Thus, there was pressure to destroy the canals to enable cultivation of large plots and to adjust the irrigation regime to the demands of cotton growing. Due to the absence of clear regulations or written guidelines and the disengagement of the system from environmental considerations, the "cash crop" pressures led to a slow destruction of water control in the area.

Drainage canals were filled in and the required high water ground-water level was not maintained. These changes led in the early 1980s to dramatic, possibly irreversible, damage to the peat area. Water level was lowered to a depth of a few meters, spontaneous fires spread, the dry soil became hydrophobic and did not hold water, and soil subsidence was accelerated. Soil fertility was reduced and parts of the area became desert.

Decision Process

The conventionally accepted response to land subsidence was to deepen the drainage canals to prevent the flooding of low-lying plots. The Hula Committee decided in 1982 not to continue this response but rather to reevaluate the situation. The committee decided to assess all possible alternatives for the management of this area. The first alternative to be considered was to continue the emphasis on the conventional farming system. This alternative would require the deepening of the drainage system together with an effort to solve the soil fertility problems. A second major alternative was to reflood part of the area, to solve environmental problems as well as to create sources of income from tourism and aquaculture. This option raised widespread objections from the agricultural lobby, mostly due to the fear that land rights would be taken if the area were reflooded.

The Hula Committee, together with the Lake Kinneret Authority, guided a prefeasibility study of the different options. One guiding principle was that the evaluation should be based, in the first phase, only on economic considerations. The finding of the prefeasibility study was clear: flooding and expansion of tourism in the area was not only an environmentally friendly option, but is the one that would generate more income. The recommendation of the Hula Committee to reflood part of the valley came at a time when the income from farming was risky and meager and when tourism was expanding. These trends encouraged the acceptance of the proposed project by the people in the region and by the different authorities. Presently, about 200 ha of previous peatlands are covered by water, forming an artificial lake surrounded by an area of wetlands. Water level in the area is maintained at a high level, and a number of projects are being planned.

BOX 4.2 (continued)

Evaluation of the Decision Processes

The Decision to Drain: 1940-1950

Had the initial planning process for drainage of the lake included a serious environmental impact statement (EIS), it is quite probable that the decision would still have been to drain the lake. Most present-day institutions and experts would approve the project considering the information available and the external conditions at the time.* However, an EIS would have indicated the weak points of the project. It is possible that it would have directed the planners to better preserve the natural assets. Better ways might have been found to manage the area, especially in the management of ground-water levels to help minimize land subsidence and decomposition of the organic soils.

We do not have the ability to predict environmental implications of human modifications of natural systems. One role of an EIS is to present a range of possible environmental implications. Such a list should direct a monitoring system aimed at getting an early warning of negative environmental developments. The Hula drainage project did not include any monitoring effort. Thus, changes in the environment were "discovered" only 15 to 20 years following the drainage when the changes were partially irreversible and almost catastrophic.

A clear response system should be ready in case negative environmental developments are detected. Such a system was not prepared and is often missing in environmental management processes worldwide. A project is approved by the environmental authorities following the development of a given set of predictions and expectations that show the effects of the project on the environment (the external effects) are acceptable. The ability to change the decision when the predictions and expectations are proved wrong is limited. By that time, investments have been made, and economical, political, and social interests are involved; and it is difficult to demand major modifications.

This is not the case when changes in conditions affect the internal functioning of a given project (e.g., the development of a better alternative product by a competing industry). Such a conditional approval system is needed to respond to unforeseen environmental effects of water resource development. An environmental approval should be reviewed periodically (e.g., once every 10 years), and the project owner must take some risk in case the project is found to cause negative environmental impacts. In the case of the Hula project, such conditional approval was not properly formulated and implemented.

---

*There was some opposition to the project by the "greens" of the time. They included persons from the region who enjoyed the area as it was and did not see it as a menace. Opposition to drainage was the driving force in establishing the first and strongest environmental nongovernmental organization in Israel, the Society for the Protection of Nature.

features of the downstream flow incrementally and in various ways, so long-term studies are needed to evaluate dams' effects and to determine how much water should be released and how often, to reduce their damage to downstream biodiversity.

Indirect Effects of Water Use on Biodiversity, Ecosystem Services, and Water Quality

Water made unavailable to support biota at one site is usually transported to another site to support agricultural production or urban development. Year-round agriculture in the study area's drylands is totally dependent on irrigation, and neither agriculture nor urban areas can depend on local water resources. Water-resource development thus is a prerequisite for both agricultural and urban development. The dramatic increase in the extent of agricultural land and of urbanization in the study area, then, obviously reflects a significant increase in water use.

Although both agricultural and urban development require water-resource development, agriculture uses more land and has more effect on biodiversity than urban development. Agriculture, urban development, and the infrastructure connecting them (e.g., roads and pipelines) adversely affect biodiversity in two ways: through the loss of natural ecosystems by land transformation, and the loss of biodiversity in natural ecosystems resulting from their fragmentation, especially by infrastructure. Two other effects are restricted to agriculture: the damage to species in adjacent and nearby ecosystems caused by airborne pesticides, and the contamination of aquatic ecosystems and aquifers by pesticide and fertilizer runoff.

Loss of Natural Ecosystems and Biodiversity

The millions of dunams of agricultural land in the study area, much of it under intensive cultivation, means the loss of millions of dunams of natural ecosystems. The specific contribution of these lands to aquifer recharge and how it may have changed since their transformation to agricultural lands is not known. But it is likely that at least some damage has occurred. In fact, such indirect effects of water use on natural ecosystems may be the most undesirable effects on water supplies, more undesirable than the direct effects of water use on natural ecosystems. The dimensions of the loss of this service (recharge) depend on the geomorphological properties of the transformed watershed, its geographical placement with respect to regional aquifers, the diversity of ecosystem types in the landscape, and the type and structure of its natural vegetation. The degree of loss also depends on the properties of the development, that is, on the agrotechnological practices and type of crops.

The allocation of land for agriculture and urban development has not taken such issues into account in the study area and most regions of the world. Evaluating the amount of water lost due to the appropriation of natural watersheds by agriculture and urban development in Israel, the West Bank and Gaza Strip, and Jordan is important. Such study may guide priorities for land uses within countries of the study area, keeping in mind that ecosystem services can be restored when agricultural land is abandoned and the natural ecosystem is rehabilitated.

The reduction of natural ecosystems also causes local extinction of populations and species, that is, reduction of biodiversity, regardless of the loss of ecosystem services. The persistence of a population is a function of its size, among other things. Population size is often a function of the area available for that population. The chances of extinction of a population dramatically increase when its available area, hence its size, is reduced below a species-specific threshold (NRC, 1995b). Similarly, in general, as the area of a natural ecosystem decreases below an ecosystem-specific threshold, the number of species inhabiting it decreases (Soulé, 1986); in general, species richness is positively correlated with habitat area (McArthur and Wilson, 1967). In this case too, it is not known how many species' populations have been lost in the study area through reduction in natural ecosystem size, by transforming these ecosystems via water resource developments.

Although Israel has thus far (1998) lost only one mammal, one frog, and one fern from its aquatic and riparian biota, many more species are at high risk, especially amphibians (Yom-Tov and Mendelssohn, 1988). Moskin (1992) divided the 491 mammal, reptile, amphibian fish, fern, and monocotyledonous plants (excluding grass) species of Israel into two categories: aquatic or riparian versus nonaquatic. In each category, he compared the number of species at risk of extinction using the International Union for the Conservation of Nature (IUCN) categories of "extinct," "endangered," and "vulnerable" with those not at risk of extinction (using their categories "rare," "insufficiently known," and "out of danger").

Whereas only 14 percent of nonaquatic species were at risk, 35 percent of aquatic species were at risk, representing a statistically significant difference that was seen in each of the taxonomic groups. For fish (only aquatic, of course), the proportion of species at risk was similar to those in nonaquatic species in other groups. Nathan et al. (1996) showed that, although waterfowl and raptors consist of only one-third of the regularly breeding birds of Israel, all but one of the 14 extinct bird species of Israel were waterfowl (7 species) or raptors (6 species, 4 of which were mostly wetland or riparian). These data suggest that further reduction in the size or water quality of aquatic ecosystems in the area could cause the extirpation of more than 35 percent of their vertebrate and plant species (and probably a high number of invertebrate species as well).

It is not known which regional ecosystems are more prone to species loss through reduction in size, nor what their thresholds are below which species are lost (see Appendix C). Thus, the fact that the number of extinctions in terrestrial ecosystems of the study area known so far is insignificant is not a cause for complacency. To conclude, the appropriation of land by agricultural and urban development impairs at least one water-related ecosystem service—recharge—and also jeopardizes regional biodiversity.

A troubling example of the loss of biodiversity is the loss of natural ecosystems in the Negev desert. In the 1950s, Israel promoted "greening the desert," resulting in a transformation of traditional rangeland to irrigated cropland, adversely affecting peripheral populations of plants and animals of rich within-species (genetic) diversity (Safriel et al., 1994). Increased ubanization, technological advances in wastewater treatment, recognition that agriculture in the central coastal plain endangers the coastal aquifer, and irrigation via the Israeli National Water Carrier all encourage this shift of Israeli agriculture from relatively wet, fertile regions to semi-arid regions. But this shift accelerates the loss of biodiversity, and probably the provision of some ecosystem services. Thus, it is not sustainable over the long term. The loss of natural ecosystems and biodiversity occasioned through Israel's policy of greening the desert gives cause for concern about the potential adverse effects of Jordan's Badia Program to develop its eastern desert.

Effects of Fragmentation. An agricultural plot that dissects a natural ecosystem, or even a road cutting through that ecosystem, can split a large and hence safe ecosystem into two smaller, extinction-prone ones. Migration between two small ecosystems can offset the risk of species extinction in each, at least in any ecosystem that functions as a sink for migrants from another. But the development that caused the fragmentation often serves as a barrier for migration. Similarly, this barrier can erode within-species genetic variability, further contributing to risks of species extinction (Tilman et al., 1994). Statistics on road casualties of endangered species suggest that roads function as effective barriers between ecosystems. But the effects of fragmentation on the study area's biodiversity has not been studied.

Effects of Pesticides. Pesticides are applied rather generously in the study area; for example, 15,000 metric tons are applied every year in Israel. Especially when applied from the air, the effect of pesticides on natural ecosystems adjacent to agricultural land is evident. Pesticides and herbicides are often concentrated at each link of food webs, sometimes at up to lethal concentrations in top trophic levels. Top-down effects on ecosystems may be highly significant, hence pesticides cause a great concern. Pesticides are also transported by runoff affecting aquatic ecosystems. The recent reductions in cotton production in Israel, for example, not only save water, but also reduce the pesticide damage to aquatic and other ecosystems.

Effects of Fertilizers. Fertilizers too are applied in large quantities in the study area, often in the irrigation water. Fertilizers reach aquatic ecosystems, where they can cause eutrophication, and they also contaminate ground water. Thus, water drawn from lakes, rivers, and aquifers for agriculture contaminates and alters ecosystem functioning. Again, such indirect effects of water use may be environmentally more significant than their direct effects. Because dryland ecosystems are limited not just by water but also by nutrients, the enrichment of fertilizers "escaping" from desert agriculture may dramatically change the functioning and structure of these ecosystems.

Effects of Trace Elements. Effects of trace elements have not received sufficient attention in the study area. However, the experience of irrigated agricultural development in the San Joaquin Valley in California (NRC, 1989) suggests that harmful trace elements, especially selenium, are abundant in agricultural drainage water, and these can be further concentrated in the food web, damaging wildlife and humans.

MITIGATION

What is being done and what should be done to mitigate adverse effects on natural ecosystems and their biodiversity, as they are caused by current and future water-resource development in the study area? Mitigation activities are of four types: restoration of damaged aquatic ecosystems; securing allocation of water for aquatic ecosystems, thus guaranteeing their ecosystem services for the future; development and implementation of a system for environmental impact assessment of planned major water-management projects in the study area; and development of regional planning policies that integrate water-resource development, agricultural development, and the functioning of natural ecosystems, to promote overall sustainability.

Wastewater for Restoration of Freshwater Ecosystems

Until 1991, the prevailing notion was that aquatic ecosystems should be rehabilitated by elimination of all effluents, ensuring flow of freshwater only. But the realities of water scarcity in the study area made it clear that rivers will dry up completely if the discharge of high-quality effluents back to them is not permitted when freshwater allocations are unavailable. For example, the Hula Nature Reserve in Israel has been found to function even when much of its water is effluent. The notion of using wastewater to help support biodiversity is based also on the belief that natural ecosystem can "serve themselves" by processing the wastewater. Many data have been accumulated, for example, along the course of the Yarkon River, to evaluate the treatment capacity of this river. For the month of June 1994, self-purification during the passage of water through measured sections of the river was evident in reductions of 0.1 to 0.5, 0.5 to 0.6 and 0.2 microgram/liter/second respectively in biological oxygen demand, chemical oxygen demand, and ammonium concentration—a high rate of self-purification, typical of an eastern Mediterranean climate (Rahamimov, 1996). Similar values have been measured in the plains section of the Soreq stream, and much higher values in the mountainous sections of this sewage stream. To increase the self-purification potential of the Yarkon, small dams have been constructed and the slowed-down stream above them is artificially oxygenated. An Israeli National River Administration was established in 1993 and charged with coordinating the restoration of river ecosystems, including the use of wastewater for this purpose. Though the main motivation for such action is recreation, the rehabilitated rivers promote biodiversity and provide ecosystem services. These restorations require water allocation of wastewater of specified quality, as well as freshwater allocation. This freshwater is not necessarily water lost to agriculture, because most of the allocation can be impounded at the lower reaches of the rivers, and the fraction lost by seepage recharges aquifers.

Balancing Water Resource Development with Biodiversity and Ecosystem Services

Regional Planning Using Advanced Technologies

Intensifying water-resource development puts the study area's biodiversity and ecosystem services at risk. It is therefore necessary to evaluate the benefits of the development against the lost biodiversity and services. The risk of loss can be reduced by striking an optimal balance between land allocation for development and for biodiversity. Remote-sensing and geographic information systems (GIS) technologies are now available to carry out this mission by means of the following steps.

1. Taking stock of current land uses, classed by development (e.g., urban areas, industrial areas, rural settlements, agriculture, and infrastructure) and biodiversity (e.g., protected areas, open areas not legally protected, rangelands, and some types of extensive agriculture). Thus, the first GIS map layer can plot current development and existing biodiversity.

2. Ranking the various types of existing biodiversity (e.g., an indigenous woodland of a given successional state or of a semiarid watershed) in terms of ecological value–i.e., provision of ecosystem goods and services and support of biodiversity–and the different sizes of each of these types.

3. Assessing the relative value of existing biodiversity areas identified in step 1 using the rankings obtained in step 2. For highly developed sections of the study area, the biodiversity areas will be scattered patches of natural ecosystems within a matrix of development, with the size of each patch and its distance from adjacent patches contributing significantly to its relative value. In nondeveloped areas, patches of development will be interspersed within a matrix of natural ecosystems, and the relative value of each type of patch will be less affected by size and distances to similar patches. Relative values can be expressed as colors or color tones in a second GIS map layer.

4. Estimating the dimensions and identifying the areas required for additional, forecasted water-driven development. The economic benefits of water-resource development of each of these areas can then be assessed and expressed in a third GIS layer.

5. Overlaying the third map layer on the second layer is the first step in an iterative process leading to optimization. Given that biodiversity areas cannot be recreated, optimization will entail adjusting the development areas such that, for example, low-benefits development areas will not be overlain on high-value diversity areas. The optimization process, though, may be more complicated than just that.

The major undertaking is step 2 above, namely the ranking of biodiversity and ecosystem services. This ranking has never been done in the study area in an objective, methodical manner, and ideally should be preceded by sufficient research. However, this fact should not discourage carrying out the exercise in current and future planning. Demand simply grows faster than the pace of the required research. It is therefore necessary to use existing knowledge, and improve the valuation as knowledge accumulates.

Evaluation of Terrestrial Biodiversity and Ecosystem Services

The biota of an area can be evaluated by three criteria: its ability to provide ecosystem services; the number of species of realized and potential direct economic benefit that it includes; and its ability to absorb anthropogenic disturbances without loss in its ecosystem services or biodiversity (resistance), along with its potential for rehabilitation following disturbance (resilience; Safriel, 1987). Each of these criteria can be quantified by applying current knowledge, paradigms, or prevailing notions, as follows.

Provision of Ecosystem Services. Water-related ecosystem services depend on the property of the ecosystem and its placement within the watershed. Concerning properties, working hypotheses are that the larger the number of vegetation layers, the greater is the infiltration potential and the smaller the risk of soil erosion and intense surface runoff; and the larger the number of species, the greater the number of vegetation layers.

Although the exact number of species in most of the study area's ecosystems is not known, these ecosystems can still be ranked in species richness. Woodlands and scrublands, for example, are richer than rangelands in the number of their perennial species (in dry subhumid areas), and stabilized sand dunes are richer than salt pans (in semiarid and arid areas). Vegetation maps that depict the major plant formations, such as those just mentioned, are available for most parts of the study area (Zohary, 1973), and the numbers of their species are also available in various sources. It is therefore possible to rank all of these major plant formations of the area by their number of plant species.

With respect to the placement of the ecosystem within the watershed, the higher the elevation of an ecosystem within the watershed, the greater the value of its services. For example, loss of woodland at the top of a watershed, where rainfall in the area is more abundant, will generate more destructive floods, with a greater loss to aquifers, than similar loss at the bottom of the watershed. Ecosystems can therefore also be scored according to their elevation above the bottom of the watershed.

Species of Potential Economic Value. An ecosystem with a large number of species is also likely to have a relatively large number of species of potential economic significance. An ecosystem can be ranked by its number of species not only to evaluate its biodiversity, but also to assess the potential economic value of its biodiversity. Sometimes, it is even possible to identify particular species whose potential is already realized. The following groups of species can be ranked by their realized or potential economic value, the top rank being most valuable: (1) progenitors of cultivated species; (2) wild relatives of cultivated species; (3) noncultivated species currently collected for nutritional, medicinal, ornamental, aromatic, energy production, and industrial purposes; (4) high-quality forage species; (5) low-quality forage species; (6) species represented by peripheral populations; (7) species already identified by IUCN revised criteria under the categories of vulnerable and rare (including species whose economic significance is not yet known, but whose extinction would prevent the discovery of their significance); (8) species of inspirational and recreational value (which often translate to economic benefits); (9) species of scientific interest (which also have economic value, including through scientific discoveries); and (10) species that provide or manipulate habitats for other species, or are ecosystem engineers (Jones et al., 1994). An ecosystem can be scored by the number of its species in each of the above categories, multiplied by the rank of the category.

Resistance and Resilience. Resilience and resistance are positively (but not linearly) correlated with area. The risk of extinction is reduced with greater population size, population size increases with area, number of species increases with area, and the large perimeter-to-surface ratio of small areas makes their species highly vulnerable to surrounding development. However, it is difficult to prescribe the threshold size for an area to be nonresistant or nonresilient. Hence, in the study area, which as a whole is small, the larger the area allotted to natural ecosystems, the better.

Rehabilitation of biodiversity and ecosystem services following disturbance is faster when there are sources of immigrants. These sources are other natural ecosystems, so their significance increases as they are closer to the disturbed area. The penetrability of the surrounding areas for propagules interacts with their distance: the greater the penetrability of the areas, the farther the propagules can travel. For example, for many species, an extensive surrounding agricultural area is more penetrable than a surrounding urban region.

To conclude, the most valuable ecosystem is one with highest number of species, many of which are of potential economic significance; one that performs unusual or particularly valuable services; and a large ecosystem, especially if it is connected by a corridor to another similar natural ecosystem.

Evaluation of Aquatic Biodiversity

The study area is relatively poor in aquatic ecosystems. Therefore, in evaluating biodiversity, a higher score should be attributed to areas that contain aquatic ecosystems, or to each aspect of an aquatic ecosystem, than to a terrestrial ecosystem otherwise having the same scores. Thus, the value of an aquatic ecosystem in the study area with a given number of species will be higher than that of a terrestrial ecosystem of the same number of species and the same size. The following identifies some guidelines for evaluating aquatic ecosystems.

In their provision of ecosystem services and number of species, aquatic ecosystems can be ranked as follows from greatest to less great: lakes, wetlands, ephemeral ponds, springs, perennial rivers, and streams. The higher the elevation of an aquatic ecosystem within a watershed, the greater the value of its services. Aquatic ecosystems also affect biodiversity of adjacent terrestrial ecosystems, by providing water for terrestrial vegetation, and water and food for terrestrial fauna.

With respect to species' economic value, the category of forage species in the previous list of terrestrial ecosystems should be replaced by species of fisheries significance. Special features of aquatic ecosystems that confer resistance and resilience, apart from features described for terrestrial ecosystems, are the distance of the ecosystem from polluting sources, which should be great, and the existence of corridors, such as streams, between isolated water bodies.

Using these sets of rules, it should be possible to evaluate regional biodiversity, and to use this evaluation as a tool to determine the extent of desirable water-resource development, such that this development is sustainable. Even if knowledge is incomplete, any serious attempt to rank areas in this fashion is likely to lead to better decisions.

RECOMMENDATIONS

This chapter has shown that maintaining and enhancing ecosystem goods and services will help—not hinder—most aspects of economic development and welfare in the study area. These goods and services enhance the quality of life of the study area's inhabitants; and they are required to maintain environmental quality, including water quality. The chapter has shown that biological diversity is important as well, and protecting it is likely to protect the structure and functioning of ecosystems to achieve those benefits; maintaining ecosystem goods and services will also protect biodiversity. The two points above require that, in plans for providing and allocating water resources among various uses in the study area, a balance is needed among environmental, economic, and other objectives when they do not lead to the same priorities for water use.

Two types of recommendations follow. The first outlines the scientific information needed to better understand the relationships among ecosystem goods and services, ecosystem structure and functioning, and biodiversity, and also the information needed to assess the balances and tradeoffs among various objectives. The second set of recommendations outlines ecologically based methods for improving the sustainability of water supplies, based on scientific information already in hand.

Research Recommendations

1. Identify and quantify the services provided by each of the study area's ecosystem types, distinguishing between water-related services, and other services. Study and quantify the optimal and minimal water allocations (quantity and quality, in time and space) for each of these ecosystems to sustain the provision of each of their services.

2. Determine which of the ecosystem types within the study area's landscapes play landscape-relevant keystone roles and investigate ways to maintain natural processes, and hence diversity at the landscape and region scale, while meeting the human demands of these landscapes.

3. Identify species of the study area that are endangered or at risk of becoming endangered, assess the contribution of each to water-related ecosystem services, identify the causes for the endangerment of these species, and explore means to reduce the risks.

4. Compare local water losses from evapotranspiration of natural and moderately managed major ecosystems of the study area to regional water gains from each of the ecosystem services, including increasing infiltration and reducing surface runoff and its associated topsoil erosion.

5. Assess the study area's biodiversity components (species, ecotypes, and populations) of current and potential economic significance, especially in aquatic habitats and climatic transition zones inhabited by peripheral populations, and determine the water allocation and the land area and configuration required for their conservation.

6. Assess the economic and biodiversity significance of the study area's indigenous dryland trees, especially the desert acacia, and the effects of current and potential relevant development projects (wells, dams, and roads) on the sustainability of the trees.

7. Conduct long-term studies to evaluate the effects of damming storm water on biodiversity at the lower reaches of watersheds, especially in hyperarid and arid regions, and use the results to prescribe amounts of water that must be released to reduce damages to downstream biodiversity.

8. Evaluate the amount of water lost through regional appropriation of natural watersheds by agriculture and urban development, to generate guidelines for land use allocation in areas still not developed and for changes in current land use.

9. Study the rate of extinctions of species populations in the study area resulting from fragmentation, transformation, and reductions in size

of natural ecosystems, and use the results to provide guidelines for water management and related development projects.

10. Evaluate the amounts of water allocated to nature reserves and other ways of protecting biodiversity that go to recharging aquifers after these uses.

11. Study the role of the area's natural ecosystems in treating wastewater of various quality, the degree to which freshwater allocated to natural ecosystems can be replaced by treated wastewater, and the technologies appropriate for this substitution.

12. Conduct the research required to define improved criteria for evaluating the significance of the area's biodiversity in providing ecosystem goods and services.

Operational Recommendations

1. The sustainability of water supplies requires that the area's natural ecosystems be treated as one of the legitimate users of the study area's water resources.

2. Because water-resource development and the further development it promotes can damage biodiversity and therefore impair the provision of ecosystem services, development in the study area should be carried out so that the gains of water-resource development clearly outweigh lost ecosystem services and reduced biodiversity.

3. Precise objectives should be set for all aquatic, riparian, and other water-dependent sites in the study area, specifying the type of biodiversity to be maintained and the type of ecosystem service the site can provide and whose continuance should be ensured. These objectives should be used to determine the minimal required allocation of water quantity and quality. Indicators, benchmarks, and monitoring programs for each water-allocation site should be developed to review and update the allocations.

4. In future land-use planning, as in water-resource planning, the benefits of proposed developments should be evaluated against the cost of lost biodiversity and reduction of ecosystem services.

5. When the study area's climatic transition areas (rich in within-species or genetic diversity), as well as other areas rich in progenitors and relatives of domestic crops, are targeted for water-driven development, it would be prudent to consider setting aside within them protected areas sufficiently large to serve as repositories of genetic resources.

6. The costs and benefits of avoiding, reducing, or mitigating the effects of fragmentation of natural ecosystems should be considered when planning water development and allocation and the additional development they promote.

REFERENCES

Anonymous. 1992. UN Convention on Biological Diversity. Geneva, Switzerland: United Nations Environmental Program

Ashkenazi, S. 1995. Acacia trees in the Negev and the Arava, Israel. Leisrael, Jerusalem: Hakeren Hakayemet (Hebrew with English summary).

Ben-David, Z. 1987. Tanninim River, nearly the end of the road. Tel Aviv, Israel: The Society for the Protection of Nature in Israel, report (in Hebrew).

Boeken, B., and M. Shachak. 1994. Desert plant communities in human-made patches, implications for management. Ecological Applications 4:702-716.

Carmel, Y., and U. Safriel. 1998. Habitat use by bats in a Mediterranean ecosystem in Israel, conservation implications. Biological Conservation 84:245-250.

Cowardin, L. M., V. Carter, F. C. Golet, and E. T. LaRoe. 1979. Classification of Wetlands and Deep Water Habitats of the United States. Washington, D.C.: Office of Biological Services, Fish and Wildlife Service, U.S. Department of the Interior.

Frankel, O. H., and M. E. Soulé. 1981. Conservation and Evolution. Cambridge, U.K.: Cambridge University Press.

Gazith, A., and Y. Sidis. 1981. Report on a survey of coastal ephemeral pools 1979/80. Tel Aviv, Israel: Institute of Nature Conservation Research, Tel Aviv University.

Grime, J. P. 1997. Biodiversity and ecosystem function: The debate deepens. Science 277:1260-1261.

Hoyt, E. 1992. Conserving the wild relatives of crops. Rome, Italy: IBPGR.

Jones, C. G., J. H. Lawton, and M. Shachak. 1994. Organisms as ecosystem engineers. Oikos 69:373-386.

Lawton, J. H. 1991. Are species useful? Oikos 62:3-4.

Mador-Haim, Y. 1987. Brechat Ya'ar Pool. Teva VeAretz (Nature and Land) 29(3):45-47 (in Hebrew).

McArthur, R. H., and E. O. Wilson. 1967. The Theory of Island Biogeography. Princeton, N.J.: Princeton University Press.

Moskin, Y. 1992. The Influence of Mankind on Aquatic Ecosystems. An unpublished dissertation, submitted to the Hebrew University of Jerusalem (in Hebrew).

Nathan, R., U. N. Safriel, and H. Shirihai. 1996. Extinction and vulnerability to extinction at distribution peripheries: An analysis of the Israeli breeding avifauna. Israel Journal of Zoology 42:361-383.

National Research Council (NRC). 1989. Irrigation-Induced Water Quality Problems. Washington, D.C.: National Academy Press.

National Research Council (NRC). 1992. Restoration of Aquatic Ecosystems. Washington, D.C.: National Academy Press.

National Research Council (NRC). 1995a. Wetlands: Characteristics and Boundaries. Washington, D.C.: National Academy Press.

National Research Council (NRC). 1995b. Science and the Endangered Species Act. Washington, D.C.: National Academy Press.

National Research Council (NRC). 1997. Innovations in Ground Water and Soil Cleanup: From Concept to Commercialization. Washington, D.C.: National Academy Press.

Noy-Meir, I. 1973. Desert ecosystems: Environment and producers. Annual Review of Ecology and Systematics 4:25-51.

Perrings, C. 1991. Reserved rationality and the precautionary principle: Technological change, time and uncertainty in environmental decision making. Pp. 153-166 in Ecological Economics, R. Costanza, ed. New York: Columbia University Press.

Puigdefabregas, J. 1995. Desertification: stress beyond resilience, exploring a unifying process structure. Ambio 24:311-313.

Rahamimov, A. 1996. Master Plan for the Yarkon River. Tel Aviv, Israel: Yarkon River Authority (in Hebrew).

Raven, P. H., and G. B. Johnson. 1992. Biology. St. Louis: Mosby.

Safriel, U. N. 1987. The stability of the Negev Desert ecosystems: Why and how to investigate it. Pp. 133-144 in Progress in Desert Research, L. Berkofsky and M. G. Wurtele, eds. Totowa: Rowman and Littlefield.

Safriel, U. N., S. Volis, and S. Kark. 1994. Core and peripheral populations and global climate change. Israel Journal of Plant Sciences 42:331-345.

Sagoff, M. 1988. The Economy of the Earth. Cambridge, U.K.: Cambridge University Press.

Sagoff, M. 1996. On the value of endangered species. Environmental Management 20(6): 897-911.

Soulé, M. E., ed. 1986. Conservation Biology: The Science of Scarcity and Diversity. Sunderland, MA: Sinauer Associates.

Stanhill, G. 1993. Effects of land use on the water balance of Israel. Pp. 200-216 in Regional Implications of Future Climate Change, M. Graber, A. Cohen, and M. Magaritz, eds. Jerusalem, Israel: Israel Academy of Sciences and Humanities.

Tchernov, E. 1994. New comments on the biostratigraphy in the middle and upper pleistocene of the southern Levant. Pp. 333-350 in Late Quarternary Chronology and Paleoclimates of the Eastern Mediterranean, O. Bar-Yosef and R. S. Kra, eds., Tucson, Arizona: Radiocarbon.

Tilman, D., R. M. May, G. L. Lehman, and M. A. Nowak. 1994. Habitat destruction and the extinction debt. Nature 371:65-66.

U.S. Environmental Protection Agency (EPA). 1993. Constructed Wetlands for Wastewater Treatment and Wildlife Habitat. EPA832-R-93-005. Washington, D.C.: U.S. Environmental Protection Agency.

Ward, D., and C. Rohner. 1997. Anthropogenic causes of high mortality and low recruitment in three Acacia tree taxa in the Negev desert, Israel. Biodiversity and Conservation 6:877-893.

Yom-Tov, Y., and H. Mendelssohn. 1988. Changes in the distribution and abundance of vertebrates in Israel during the 20th Century. Pp. 515-547 in The Zoogeography of Israel, Y. Yom-Tov and E. Tchernov eds. Dordrecht: W. Junk.

Zohary, D. 1983. Wild genetic resources of crops in Israel. Israel Journal of Botany 32:97-127.

Zohary, D. 1991. Conservation of wild progenitors of cultivated plants in the Mediterranean basin. Botanika Chronika 10:487-474.

Zohary, M. 1973. Geobotanical Foundations of the Middle East. Gustav Fischer Verlag, Amsterdam, Stuttgart Swets and Zeitlinger.

5
Options for the Future: Balancing Water Demand and Water Resources

The conventional freshwater sources now available in the region are barely sufficient to maintain the study area's current quality of life and economy. Jordan, for example, is currently overexploiting its groundwater resources by about 300 million cubic meters per year (million m^3/yr), lowering water levels and salinizing freshwater aquifers; similar examples of overexploitation are occurring throughout the study area. Attempting to meet future regional demands by simply increasing withdrawals of surface and ground water will result in further unsustainable development, with depletion of freshwater resources and widespread environmental degradation. Because these conditions already exist in many parts of the study area, for example in the Azraq Basin and the Hula Valley, as described in Chapter 4, the reality of a constrained water supply must be considered in formulating government economic plans and policies. It seems likely that demand and supply can be brought into a sustainable balance only by changing and moderating the pattern of demand, or by introducing new sources of supply, or both. Above all, water losses should be minimized and water use efficiency increased substantially.

MANAGING DEMAND

Water shortages have already been faced in the study area as a result of droughts, and they have been overcome by managing demand. The reduction in Israeli water use from 1,987 million m^3/yr in 1987 to 1,420

100

million m³/yr in 1991—with no net loss in agricultural production or economic growth (Biswas et al., 1997)—indicates what can be accomplished in the way of demand moderation. In practice, demand for water can be influenced by conservation measures in urban, agricultural, and industrial sectors, and by economic (pricing) policies. It is important to recognize that, while demand management efforts may economize on water effectively, they are also rarely costless. In this section, a number of demand-management policies are described and discussed in light of the committee's five evaluation criteria (see Chapter 3 for a full statement of the criteria).

Conservation

Given the inevitability of population growth, it is imperative that per capita consumption of water in the study area be addressed through conservation measures in all three major sectors of water use: urban, agricultural, and industrial. There are significant disparities in per capita water use within the study area, and there will doubtless be pressures to raise the lowest consumption rates to parity with the highest rates. However, some middle ground must be reached to bring quality of life and economic development into balance within the practical constraints imposed by the region's available water. This balance requires lowering the study area's capita water use without significantly degrading the economy or standard of living, and at the same time improving the economy, hygienic conditions, and standard of living among Jordanians and Palestinians.

Conservation measures to reduce water demand are generally well established, but they often require societal or economic incentives to implement. Although some conservation measures are costly, most compare favorably with measures to increase water supplies. Moreover, water conservation measures invariably have a positive effect on water quality and the environment, if only by minimizing the impacts on freshwater resources and the volumes of wastewater generated by human activities.

Urban

In urban and rural-domestic sectors elsewhere, notably the United States, conservation measures are most effective when they have broad public support. Important voluntary domestic water conservation measures include the following:

- Limiting toilet flushing.
- Adopting water-saving plumbing fixtures, such as toilets and shower heads.

- Adopting water-efficient appliances (notably washing machines).
- Limiting outdoor uses of water, as by watering lawns and gardens during the evening and early morning, and washing cars on lawns and without using a hose.
- Adopting water-saving practices in commerce, such as providing water on request only in restaurants and encouraging multiday use of towels and linens in hotels.
- Repairing household leaks.
- Limiting use of garbage disposal units.

Public support of such measures is highly variable because many of them are voluntary, relying on individual actions, or have negative societal impacts (such as higher prices or taxes, as discussed below). The study area is at an advantage in this regard, because the population is relatively well aware of how water is used. Public awareness of levels of water use is the key to effective urban conservation programs.

Voluntary domestic conservation measures can result in significant water savings. Among such conservation measures, low-flush toilets use approximately 6 liters of water per flush, while conventional toilets operate with 13 to 19 liters. Toilet water-displacement devices, such as a simple water-filled plastic container, are placed in the toilet tank to reduce the amount of water used per flush. A toilet dam, one type of water-displacement device, saves 3.7 to 7.5 liters per flush. Low-flow shower heads are relatively inexpensive and save 7.5 liters per minute (U.S. EPA, 1995). Installing pressure-reducing valves can also save energy as well as water by reducing the probability of system leakage and breakdowns. The U.S. Environmental Protection Agency (U.S. EPA, 1995) estimated water savings for a house with low pressure, compared to a house with high pressure, to be 6 percent. Pressure reduction increases the reliability of water systems by 33 percent (Al-Weshah and Shaw, 1994).

According to Bargur (1993), the Israeli Water Commission has estimated that municipal water use in Israel could be reduced by 55 million m^3/yr if voluntary conservation measures were widely implemented. Table 5.1 summarizes the household water savings that can be achieved using water-saving appliances. Because of the possibly significant cost savings of these voluntary conservation measures, their widespread adoption should be encouraged in the study area. As an example of the potential savings, a typical family of five persons that does not employ water conservation measures uses about 42 m^3 of water per month, at a cost of US$42 (assuming an exchange rate of US$1 = 3.6 New Israeli Shekels [NIS] as of March 1998). With the full use of water-saving devices, monthly use for five people would likely be between 16 and 19.5 m^3, a

TABLE 5.1 Comparison of Conventional and Nonconventional Appliances in Domestic Water Use

Item	Conventional (l/unit)	Improved (l/unit)	Use Frequency (times/time/steps)	Savings with Improved Technology (m³/unit)	Savings/Month (m³)	Values (US$)
Toilet Flush	16 l/flush	Low-flow (6 l/flush)	5/person/day	0.050/day	1.5	1.5
		Displacement device 8.7-12.3 l/flush		0.020/day	0.60	0.60
Shower Head	18.7-30 l/min at full capacity	Pressurized low-flow 7.5 l/min	1/person/day (10 min)	0.112-0.337/10 min	3.38-6.75	3.38-6.75
Bath	94-131 l/day					
Tap	12-19 l/min at full capacity					
Washing Machine	Full automatic, 131-263 l/wash	Manual 40-60 l/wash	3 cycles/wash	0.273-0.609/wash	0.588	0.588
Car Wash	Normal hoses for 20 minutes, 375 l	Pressurized hose for 20 minutes, 56 l	3/month	0.319/wash	0.96	0.96

SOURCE: U.S. EPA, 1995.

reduction of as much as 62 percent. This amount could be a highly significant savings, especially for low-income families in the study area.

Involuntary conservation measures applied to the urban sector are easier but more expensive to implement. Such measures include repairing leaking distribution systems and sewer pipes, expanding central sewage systems, metering all water connections, and rationing and water use restrictions. Improvements to municipal water systems, such as repairing or replacing leaking distribution systems, and achieving total metering of all water use could be accomplished as part of government policy. An aggressive conservation policy, such as the one adopted in the Mexico City Metropolitan District, can extend to the elimination of household leaks as well as to the repair of leaks in the water distribution network (NRC, 1995). These measures are expensive, but the costs and potential water savings must be weighed against the costs of developing equivalent alternate supplies. The dynamic growth possible in both public and private sectors throughout the study area holds the promise of incorporating water conservation measures into the new infrastructures to be built.

Although per capita domestic water use in Israel has been increasing, from 80 cubic meters per year (m^3/yr) in 1965 to 100 m^3/yr in 1995, the disparity among localities in per capita use suggests that water use can be reduced without degrading the quality of life. While in Jerusalem the average per capita water use is 67 m^3/yr, it is 117 in Tel Aviv, and 89 in Haifa (not including conveyance losses). In the low-income municipalities, water use rates are as low as 40 m^3/yr (Tahal, 1993). On the other hand, the domestic per capita use in rural areas in Israel is 196 m^3/yr. The disparity in per capita use needs to be further investigated, insofar as it suggests that, under conditions of further constraint, there is still room for water conservation in the urban sector. Per capita water use for urban Palestinians reaches a maximum of 100 m^3/yr, similar to Israeli use, and can reach 200 m^3/yr, whereas in rural areas it is about 20 m^3/yr, reflecting the widespread unavailability of water distribution networks as well as restricted water availability in these areas. Although Palestinian urban use may be lowered through conservation, rural use is likely to increase as water distribution systems become more widespread with improvements in the level of living. This development of new and larger water distribution facilities will increase the rate of water use unless restrictions are put in place first.

As a further indication of potential water savings, municipal authorities in the West Bank report that water losses unaccounted for in the distribution network range from 26 percent in Ramallah to 55 percent in Hebron (Hebron Municipal Water Engineer, personal interview, 1996). In Jordan, average water loss is 50 percent (Water Authority of Jordan open files, 1996). The average 1990 water loss in the 40 largest municipalities in Israel was 11.3 percent (Tahal, 1993). Commonly, much of the apparent conveyance loss in municipal water systems is actually the result of illegal or nonmetered connections, or errors in metering. Accurate metering at all connections will promote the adoption of voluntary conservation measures as well as quantifying actual conveyance losses. Water leaking from freshwater distribution pipes may not be entirely wasted, because it may infiltrate and recharge ground water. On the other hand, water leaking from sewers and effluent from cesspools and other untreated waste-disposal systems pollute underlying ground water. Minimizing these sources would result in increased flows to wastewater treatment plants, which in turn would increase the amount of water available for reuse (see "Wastewater Reclamation" below).

The quantity of additional water that can be made available for reuse by reducing losses from sewers may be significant. For example, the official figure describing the total quantity of wastewater in Israel (Table 5.2) is 374 million m^3/yr. However, urban water use is 546 million m^3/yr and industrial use is about 130 million m^3/yr. About 10 to 20 percent of urban water goes to nonreturnable, consumptive uses (mostly irrigation of private and public gardens) and as much as 50 percent of industrial use is consumed. According to these assumptions, the potential amount of returned sewage from urban use in Israel is 437 million m^3/yr (assuming a consumption rate of 20 percent), and the total amount of wastewater, including industrial wastewater, should be about 500 million m^3/yr. The difference of about 125 million m^3/yr between potential and actual wastewater represents the amount of water that can be made available through reuse of wastewater if sewering were total and losses minimized.

TABLE 5.2 Collection, Treatment, and Utilization of Wastewater Effluent in Israel, 1994

District	Population	Total Wastewater	Sewered	Treated	Utilized
Jerusalem	643,267	35,412	34,596	1,957	25,994
Northern	933,448	63,828	55,312	42,461	30,068
Haifa	716,460	52,184	49,120	43,900	36,431
Central	1,170,824	78,827	72,487	70,250	29,177
Tel Aviv	1,140,523	76,747	76,628	76,628	94,884
Southern	701,330	67,262	62,628	62,372	35,974
Total	5,305,852	374,303	350,771	297,570	252,529

NOTE: All values except population are million m^3/yr.

SOURCE: Eitan, 1995.

There are several possible explanations for this high deviation between the amount of water entering the city (urban water use) and that leaving (wastewater). Again, many municipal water-distribution systems leak. An additional and significant loss of water occurs in the wastewater disposal system. In small and medium-sized communities, many homes are not connected to sewers and dispose of their wastewater into septic tanks. The amount of sewage disposed of in septic tanks in Israel, which has the most extensive sewage collection system in the study area, was estimated to be about 50 million m^3/yr. Beyond the fact that this water is not available for reuse as treated wastewater, septic tank effluent is a major cause of ground-water pollution. Another cause of water loss is the leakage of wastewater from sewers. Presently, municipalities do not have an incentive to fix leaks in sewers or to enforce closure of septic tanks. Often, the tacit interest of the municipality may be to minimize the amount of water reaching the wastewater treatment plant, to save water treatment expenses. Structural changes and improved maintenance of water distribution, and particularly wastewater distribution systems, may appear to be costly, but may be cost-effective compared to other measures that can produce comparable quantities of water.

In conclusion, Table 5.3 shows that urban water conservation efforts are attractive when evaluated against the five criteria established by the committee. Although conservation will not usually result in augmentation of available supplies—one possible exception being the repair of leaky distribution systems—conservation measures are generally technically and economically feasible; have no adverse environmental consequences; and, by conserving current water supplies, tend to preserve the resources available for both present and future generations.

TABLE 5.3 Demand Management

Committee Criterion	Conservation			
	Urban	Agriculture	Industry	Pricing
1. Impact on Available Water Supply	0	0	0	0
2. Technically Feasible	+	+	+	+
3. Environmental Impact	0	+/−	+/0	+/−
4. Economically Feasible	+	+/−	+/−	+/−
5. Implications for Intergenerational Equity	+	+/0	+	+

NOTE: + indicates positive effects, − indicates negative effects, and 0 indicates no impact.

Agriculture

Water use in the agricultural sector throughout the study area is highly controlled by government agencies, and conservation measures have proven to be highly effective in reducing agricultural water use. The reduction in Israeli water use of more than 200 million m³/yr between 1985 and 1993 was accomplished almost entirely in the agricultural sector through the use of improved irrigation methods and water delivery restrictions. Agricultural water use may become even more efficient through rationing, research, and possibly economic policy involving changes in crops. However, as regional nonagricultural water demand increases and the cost of additional water supplies becomes more expensive, the role of agriculture in the area's economy will have to be reevaluated, to conserve as much water as possible. One possibility is that the area adopt agricultural practices more in harmony with the ecological realities of drylands. Drylands are and will probably always remain marginal for subsistence agriculture, unless it is heavily subsidized by water drawn from elsewhere. The alternative for sustainability is to develop local water resources and use them prudently, and at the same time to capitalize on local conditions and local resources in producing marketable products for export. Capturing local runoff and flood water can increase water supplies for dryland extensive agriculture (Evenari et al., 1982), and reducing evaporative water loss by cropping intensively within closed environments (using "desert greenhouses") can also effectively increase supplies. The latter practice requires financial investment and innovative technologies, but it is an economical use of land and water, it avoids salinization, and it produces a high yield of exportable cash crops such as out-of-season ornamentals, fruits, vegetables, and herbal plants.

Using computer-controlled drip "fertigation" (applying fertilizer with the irrigation water) economizes on water and fertilizer use, and prevents soil salinization and ground-water pollution. Use of brackish water, often abundant in the study area's dryland aquifers, for irrigating salinity-tolerant crops increases the sugar contents of fruits such as tomatoes and melons, and hence their market price. Brackish water is very useful for intensive aquaculture in deserts. Finally, the use of treated local or transported wastewater in subsurface drip irrigation of orchards and forage could dramatically increase the production of the study area's drylands in a sustainable manner. In any reevaluation of the role of agriculture in the study area, the socioeconomic impacts as well as the environmental impacts of changing agricultural practices should be considered.

In Jordan, 1996 agricultural water use per hectare was approximately 6,800 m³/yr. To increase the efficiency of water use, the Jordan Valley Authority has recently converted irrigation systems to pressure pipe net-

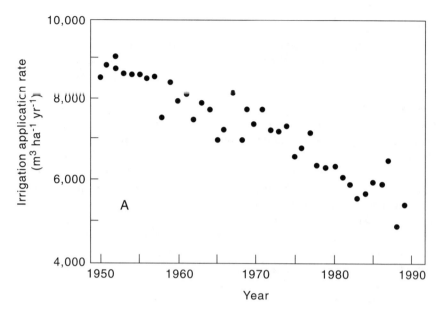

FIGURE 5.1   Average agricultural water use per hectare in Israel, 1950 to 1990.
SOURCE: Stanhill, 1992.

works. In the West Bank and Gaza Strip, 1996 agricultural water use was
about 7,150 m³/yr per hectare, much of it in drip irrigation and green-
house agriculture. In Israel, where drip irrigation is widely practiced, the
1995 average agricultural water use per hectare was 5,700 m³/yr, down
from 8,600 m³/yr in 1955 (see Figure 5.1) while crop productivity per unit
of water (see Figure 5.2) increased more than twofold, from 1.2 to 2.5
kilograms per cubic meter (kg/m³) (Stanhill, 1992).

Experience in Israel has demonstrated how water-use efficiency can
be increased by improvements in irrigation efficiency, increased crop productivity, and changes in the types of crops grown. Freshwater can also
be saved by switching to irrigation with treated wastewater or brackish
water (discussed further below). However, in Jordan the quality of wastewater effluent for irrigation may not be as high as in Israel (Salameh and
Bannayan, 1993).

At present, over 80 percent of the irrigated area in Israel uses micro-
irrigation techniques (drip and mini-sprinkler), with an irrigation efficiency of 85 to 90 percent. The remaining irrigated area uses sprinklers
with an irrigation efficiency of 75 to 80 percent. Gravity irrigation, which
has an efficiency of 50 to 60 percent, has not been used in Israel since the
mid-1960s. Automation in irrigation has resulted in better water control

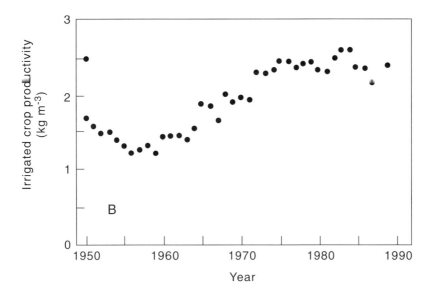

FIGURE 5.2 Irrigated crop productivity in Israel from 1950 to 1990. SOURCE: Stanhill, 1992.

and the ability to irrigate at will (to avoid windy periods, for example), thereby reducing water losses (Box 5.1).

Determining crop water requirements for different areas in Israel was a high priority of agricultural research in the 1960s and 1970s. Table 5.4 presents results of scores of experiments throughout the country for some major crops (Shalhevet et al., 1981). The Israeli government (Water Commissioner) used the results of these experiments to set water allocations for growers in the various areas.

Another way of saving water in agriculture is by shifting production from crops with high water needs to crops with lower ones. This process occurred in Israel during the early years of crop irrigation and was partly responsible for reducing average per hectare water use over the past four decades, as illustrated in Figure 5.1. There is a limit to how much this approach can contribute to water savings, however, because water quality already limits the kinds of crops that can be profitably grown in the study area.

In designing agricultural water-conservation programs, care must be taken to ensure that the water conserved is not deep percolation water that would otherwise recharge an aquifer or runoff water that supplies another person or activity. Agricultural water conservation results in a net saving of water only when the water saved would otherwise be lost

BOX 5.1 Improved Water Use Efficiency in Traditional Farming in the Jiftlik Valley[1]

Background

The Jiftlik Valley is situated in the West Bank, east of Nablus. The valley leads from the city of Nablus to the Jordan Rift Valley. Before the early 1970s, the main crops grown were winter vegetables using traditional methods, by about 4,000 tenant farmers living in six villages, with an average annual per capita income lower than U.S.$200. The landlords provided the land, water-distribution system, water-use rights, access to credit, and marketing facilities. The farmers provided labor, traditional inputs like manure, and storage and packing facilities. The farm income was split half and half. During the winter growing season, the labor requirements were high, putting a strict limit on the land area each family could cultivate. No farming was carried out during the hot summer months.

The source of irrigation water was a number of springs flowing out of the highlands to the west. The water entered earthen ditches and was led through concrete-lined canals to the fields, where it was spread by gravity methods. Average irrigation efficiency was lower than 30 percent. Water was allocated on a time basis, on a 5- to 8-day cycle, and was used by one to four consumers at a time. There were no storage facilities, since under the traditional farming and irrigation system none were needed.

The Change from Traditional to Modern Farming

Modern irrigation farming is capital-intensive, requiring expensive inputs. Thus, the first requirement is a source of capital investment. For Jiftlik, the initial investment was made possible by a loan from the Mennonite Church. Further development was made possible based on income from farm activities. In other situations, government or banks may provide the initial credit.

The change from traditional to modern farming included a package of five inputs:

1. Small farm ponds of 1,000 to 5,000 m^3 each were constructed to make water supply more dependable.

2. The traditional gravity irrigation method was replaced by drip irrigation, including peripheral equipment such as plastic supply lines, fittings, and valves.

3. Traditional crop varieties were replaced by seeds and seedlings of improved varieties, usually hybrids.

through consumptive use or severe degradation in quality. As shown in Table 5.3, agricultural water conservation will not usually increase available water. The water available does not increase because the conserved water has already been reused and allocated elsewhere. The environmental impact of agricultural water conservation also varies, with adverse impacts occurring where existing irrigation tailwaters or return flows support some environmental purpose. To the extent that agricultural

4. Plastic sheeting was used for mulching along the seedbed, construction of low tunnels during the cold season, and solar sterilization.

5. Chemical fertilizer (a compound 20:20:20 mix) was introduced, simplifying fertilizer application, along with insecticides, fungicides, and herbicides.

The farmers adopting these technologies were low income, partly illiterate, and dependent on their landlords. The landlord-tenant relationship has not been altered during the adoption of the new technologies and the lifestyle of the adult population has remained unchanged. The younger generation, however, has reaped the benefits of the improved affluence by having better health, training, and education.

The adoption of advanced technologies has led to a notable change in farming patterns. Nonirrigated crops were completely abandoned in favor of intensive vegetable cropping, fewer types of crops were grown (6 vegetable crops instead of the 23 previously grown), and land use increased by over 60 percent. It became possible to grow two crops per year, for example, fall cucumbers and winter tomatoes, or fall tomatoes and spring watermelons and melons.

From 1970 to 1986, the use of purchased inputs increased dramatically. An eightfold increase was seen in the number of tractors used and a sevenfold increase in the use of plastics, fertilizer, and seeds. The same amount of water irrigated nearly 10 times more land area than before the development process, and the irrigation method was changed from gravity irrigation (90 percent of the area in 1970) to drip irrigation (90 percent of a larger area in 1980). Yields increased three- to fivefold during the period 1965 to 1982. Equally important was the improvement in yield quality. The produce was now acceptable to the discriminating markets of the Arab oil states and Europe. Income per person increased from US$116 in 1966 to US$660 in 1974, to a peak of US$1,000 in 1980. In the late 1980s, income declined because of worsening market conditions, regional wars, and limitations imposed on imports by Jordan.

This case study demonstrates that appropriate methods and proper investments can dramatically improve the traditional inefficient irrigation methods in the Middle East region, stretching the use of available water resources, while improving the income of the farmers. At the same time, labor requirements were not lowered. On the contrary, the greater yields required more labor for harvesting, sorting, and packing, although irrigation required less labor.

---

[1]Adapted from Rymon and Or, 1991.

water conservation results in net savings of water, it will tend to preserve water resources now and for future generations.

Industry

Water use in the industrial and commercial sector accounts for a relatively small percentage of total water use in the study area, ranging from 3 percent in the West Bank to 6 percent in Israel (Table 2.3). These figures,

TABLE 5.4 Irrigation Requirements of Major Crops in Various Parts of Israel

Crop	Study Area	Irrigation Requirements (mm/year)	Expected Yield (kg/ha)
Wheat	Northern Negev	400-450	5,400
	Beit Shean Valley	500-550	5,000
Sorghum	Coastal Plain	250-280	9,900
	Northern Negev	380-420	7,800
Corn (grain)	Coastal Plain	330-350	8,500
	Northern Negev	450-480	6,800
Cotton (lint)	Coastal Plain	330-360	1,760
	Northern Negev	510-540	1,650
	Bet Shean Valley	780-810	1,700
Peanuts	Coastal Plain	530-560	5,100
	Northern Negev	530-560	4,600
Tomato (processing)	Coastal Plain	220-240	55,000
	Northern Negev	410-430	64,700

SOURCE: Shalhevet et al., 1981.

however, do not take into account the many small factories and varied commercial establishments that are supplied by municipal water systems. For these smaller users, conservation measures adopted from the domestic sector, particularly through raising public awareness and adopting water-saving technologies, can result in significant overall water savings.

Most of the larger industrial users of water are self-supplied and many, particularly in Jordan, are largely unregulated (Salameh and Bannayan, 1993). Government-imposed water-use restrictions and pricing policies, together with wastewater quality requirements and impact fees, should motivate industrial and commercial users to reduce their water use. In Jordan, the steel and paper industries are recycling their cooling water with significant water savings. Daily water demand in a steel mill was reduced from 450 m^3 to 20 m^3 per day, and in a cardboard plant from 4,000 to 800 m^3 per day (Water Authority of Jordan, written communication, 1994). In Israel, experience has shown that water recycling and the growth of industries that use very little water (for example, electronics) resulted in a decline in specific water use—that is, use of water per unit value of product. Further reductions in industrial freshwater use are possible in the study area through widespread reliance on recycling systems, the development of new water-saving technologies, the use of lower quality water, and water use monitoring.

The evaluation of industrial water conservation and pricing policies based on the committee's five criteria is summarized in Table 5.3. Industrial conservation generally does not have an effect on the quantity of available water supply because most conservation schemes involve recycling within specific industrial plants. The environmental impact of industrial water conservation can be positive when wastewater flows to the ambient environment are reduced or eliminated through increased recycling. The economic feasibility of industrial water conservation varies from industry to industry and situation to situation. Where stringent wastewater discharge regulations require that water must be treated anyway, the incremental costs of recycling the water or making it suitable for nonpotable uses may be quite attractive. Finally, industrial recycling will tend to have positive implications now and for future generations because it economizes on water use and constrains—often significantly—degradation of water quality.

Prices and Pricing Policies

Water is typically priced to serve several different and sometimes conflicting objectives. First, the purveyor has the need to cover the costs of operating and maintaining the water delivery system and the debt service on it. Second, prices can be set to ensure that water is allocated and used efficiently, so that the total costs of meeting an area's water needs are minimized and consumers economize in using scarce resources. Third, water pricing almost always involves considerations of fairness. These considerations are usually taken to mean that equals should be charged equally and water rates should be perceived as fair by water consumers. Since these objectives frequently conflict, compromises must be made among them. Depending on the circumstances, one of these objectives may be relatively more important, and pricing policies will emphasize the important objective at the expense of the others (Boland, 1993).

In most regions of the world, strategies for developing and managing water resources have focused on the provision of water supplies. The emphasis on developing and augmenting water supplies generally leads to water pricing policies that emphasize achieving revenues sufficient to cover the costs of this development. In Israel, pricing policies have frequently been established to make water readily available. Some pricing policies have been designed to induce settlement of remote lands. These policies result in prices that understate the true cost of the water, signal consumers that water is more plentiful than it really is, and require subsidies to recover costs. Similarly, pricing policies in many parts of the region emphasize affordability by setting prices low enough so that water bills remain relatively low. Water policies in the area have rarely been designed as part of an overall demand-management program and rarely result in prices that reflect the true value of the water.

It is instructive to examine policies based on the need to defray the cost of supply to illustrate how such policies may distort allocation patterns and levels of use in areas of increasing water scarcity. Water impoundment and conveyance facilities are almost always capital-intensive. This means that fixed costs are usually large compared to operating or variable costs. To ensure that the relatively large component of fixed costs is covered, pricing structures generally embody average cost-pricing rules so that the unit price becomes lower as use increases. Such rate structures tend to encourage use and discourage water economizing or conservation.

Pricing policies that emphasize economic efficiency and promote economizing in water use may be appropriate for the study area given its increasing water scarcity. Such policies will be particularly attractive when the costs of water are covered and the pricing impact of the policies conform to societal notions of fairness or equity. As a general rule, the prescription for pricing water in water-scarce regions is to set the price of water equal to the marginal cost of supplying the last unit delivered (Hirshleifer et al., 1960; Russell and Shin, 1996). As long as marginal costs are higher than average costs—which is usually the case where water is quite scarce—the use of marginal cost pricing will ensure that revenue requirements are met. Marginal cost pricing will also ensure that appropriate signals are sent to consumers about the true cost of the water and given some fixed level of benefits, will ensure that the costs of providing the water are minimized.

Time-of-use pricing, which is an amalgam of marginal cost and average cost pricing, sets the rate higher during periods of peak use to ration water during these periods, but sets lower prices during off-peak or off-season periods of uses. This kind of pricing structure tends to discourage use during peak use periods and encourage use during off-peak periods, and is particularly useful in shifting use away from peak to off-peak use periods. In regions with Mediterranean climates, time-of-use pricing might call for higher prices during the dry season, when use tends to be high, and lower prices during the wet seasons, when supplies are more plentiful and use tends to be lower (Sexton et al., 1989).

Water surcharges are frequently employed to discourage excessive use. That is, a surcharge is imposed above some specified use level to discourage additional use beyond that level. Surcharges are appropriate where water can be effectively rationed by reducing excessive use. Policies that include surcharges are frequently adopted to promote fairness.

Pricing policies that encourage conservation, including marginal cost pricing, time-of-use pricing, and water surcharges, generally work best where the quantity of water demanded is reasonably responsive to price. Thus, for example, where existing water supplies are insufficient or barely sufficient to serve basic drinking, cooking, and sanitation uses, an increase in water prices is unlikely to be effective in achieving water conservation. Conversely, where water is used in discretionary ways, as for irrigating urban landscapes or for low-valued agricultural or industrial purposes, the imposition of rate structures that approximate marginal cost pricing may lead to significant and cost-effective water conservation. Similarly, time-of-use pricing may not lead to water conservation in the aggregate, but may be effective in shifting demand use away from peak to off-peak periods, thereby avoiding the expense of constructing new facilities that would be used to serve peak period uses only.

While pricing policies alone can ensure efficient allocation of water within a service area or within a particular water-using sector, pricing policies alone may not be sufficient to ensure efficient allocation of water among the various water-using sectors. Water markets that include several sectors, and which use marginal cost prices, can be used to allocate water among the sectors in an efficient manner.

Markets have the advantage of permitting transfers of water to occur on a strictly voluntary basis. That is, sellers have an incentive to sell or lease water only when the returns they receive meet or exceed the returns they could earn by putting the water to some other use. Buyers have an incentive to participate in market exchanges only when a purchase or lease represents the least costly opportunity to obtain additional water. Market transfers of water occur when the difference between the minimum price that sellers will accept and the maximum price that buyers will offer is enough to cover any costs of transport or treatment that may be needed to effect the exchange.

Free market arrangements to transfer and reallocate water are frequently criticized for failing to take into account the legitimate interests of other parties who are not directly involved in the transactions but who do nevertheless incur costs because of the transfer (NRC, 1992). Studies of the third-party impacts of market transfers during the California drought show that, in the short run at least, such impacts are likely to be relatively modest. Moreover, in regions or locales where such impacts are significant, they can be attenuated by restricting the quantities of water that can be transferred to perhaps 15 percent of the total water available in the area (Carter et al., 1994). Other studies have led to identifying a broad range of alternatives for managing and minimizing the third-party impacts of marketlike water transfers (NRC, 1992).

Even if water markets are never developed and adopted, simulation studies of water markets can be highly useful in identifying the value of water in alternative uses and regions and in identifying additional water supply and conveyance facilities that are economically justified. One such study, now being conducted under the auspices of the Institute for Social and Economic Policy in the Middle East at Harvard University, provides a number of valuable conclusions (Fisher et al., 1996). Among the tentative conclusions of this study are (1) the value of the water of the Middle East that is in dispute is relatively modest (approximately US$125 million annually); (2) conveyance facilities will need to be built to serve the growing demands of Amman and the northern highlands of Jordan prior to 2010 to avoid a water supply crisis; (3) conveyance facilities should be built to connect the districts of the northern West Bank, and a large-scale conveyance system to take water to the interconnected West Bank districts should also be constructed; (4) the Gaza Strip should be served via an expanded connection to Israel's National Water Carrier; and (5) desalination will not be economically efficient on the Mediterranean coast until at least 2020.

Where water pricing practices are employed to ensure that water is used as efficiently as possible, the policies supporting such practices may be relatively attractive when evaluated against the committee's five criteria, as Table 5.3 shows. Although pricing practices do not augment available water supplies, they are technically feasible and, where they result in a net savings of water, tend to preserve water resources now and for future generations. The economic feasibility of pricing policies will vary from situation to situation, depending on price and income elasticities, which may vary throughout the region. The environmental implications of such policies will also have to be assessed on a case-by-case basis, since policies that result in the conservation of water that might otherwise support amenity uses (such as providing water for streams or wetlands) may have adverse impacts on the natural environment.

AUGMENTING AVAILABLE SUPPLIES

Demand management alone may not be sufficient to achieve efficient and equitable allocation of water resources in the study area. This is not to say that efforts at reducing demand are futile—new sources will be expensive and, in some cases, will furnish water of a lower quality. Demand management and supply augmentation will likely both be needed to meet the future human and environmental water requirements.

Additional regional water supplies can be obtained by using what little naturally occurring freshwater is unused (through watershed management, water harvesting, and the development of nonrenewable water), by reusing water (wastewater reclamation), by developing sources of lower water quality (marginal quality water and desalinated brackish water and seawater), by importing water from outside the study area, by transferring water within the study area (though imports and transfers are not analyzed in this report), and by attempting to increase the renewable amount of water available (cloud seeding). These options are reviewed in the following sections, along with the results of the committee's preliminary evaluations of each option based on its five criteria (see Chapter 3). Before these discussions, however, the significance of maintaining water quality is discussed to highlight the importance the committee attaches to this issue.

Maintaining Water Quality

Because the availability of adequate quantities of water in the study area is inextricably bound up with issues of water quality, maintaining water quality is discussed here. The quality of the region's water supplies has been deteriorating for some time, and continued deterioration will only make the problem of water availability worse and more expensive to solve. This trend in water quality must be reversed if there is any hope of solving the water problems of the region. Any strategy to solve the area's water problems must include components to preserve and enhance the quality of water already available.

Deterioration of the natural quality of ground and surface water has been brought about by urban, agricultural, and industrial sector activities. Some of these many human activities that affect water quality include the following:

• Discharge of inadequately treated effluent from municipal treatment plants.
  • Discharge of untreated domestic and agricultural wastes.
  • Discharge of untreated or inadequately treated industrial wastes.
  • Extraction, use, and disposal of poor-quality ground water.
  • Leachate from solid waste landfills.
  • Runoff from urban drainage.
  • Peak wastewater flows bypassing wastewater treatment plants.
  • Fertilizer and pesticide residues.
  • Saline agricultural return flows.
  • Drainage of wetlands.

It is common engineering and environmental knowledge that prevention is preferable to remediation. Industrial and municipal treatment facilities must be built to solve current pollution problems and keep up with anticipated population and industrial growth. Other measures, such as regulation of pesticides and fertilizers, and proper siting and construction of solid waste landfills, are standard water-quality protection measures that should be vigorously adopted in the study area. Total watershed management should also be adopted (see next section).

Because of the emphasis currently given to the use of reclaimed wastewater in the study area, it is of some value to look more closely at the water-quality implications of this practice. Treated municipal effluent contains many undesirable constituents. Treatment to remove these constituents is successful to varying degrees, depending on the type of treatment and also on the intended reuse of the treated effluent. In urban usage of water, biodegradable organic matter—as measured by biochemical oxygen demand (BOD), chemical oxygen demand (COD), or total organic carbon (TOC)—is added to water. Sewage treatment processes, depending on their intensity, reduce the organic matter content to almost any desired level. This level is dictated by the intended use of the effluent. Effluent may also contain trace concentrations of toxic, stable organic substances, such as pesticides and chlorinated hydrocarbons, which are considerably more difficult to remove.

Pathogenic microorganisms (bacteria and viruses) and parasitic organisms, such as protozoa and helminths, are also present in sewage effluent. Their concentration is greatly reduced during normal treatment processes. Suspended solids, including volatile and fixed solids, are present in municipal effluent, and if not adequately removed can shield microorganisms from disinfection during treatment. Disinfection with chlorine can transform organic compounds present in wastewater into chlorinated organic compounds such as chloroform. These compounds have been implicated in the development of liver, bladder, and kidney cancers. These compounds may be further enhanced if the effluent reaches potable water sources that are normally subjected to further chlorination before use.

Urban use invariably results in increased inorganic soluble salts. The principal ions picked up are sodium (Na^+), chloride (Cl^-), calcium (Ca^+), and sulfate ($SO_4^=$). These ions increase the total salt content (salinity) and the sodium adsorption ratio of the water. Unlike organic constituents, inorganic salts are not removed during conventional reclamation processes. Irrigation with water enriched in inorganic salts results in soil salinization and increased salinity of underlying aquifers and surface waters. Thus, irrigation, like many other uses of water, degrades water quality for later uses. Municipal effluent also contains increased quantities of nitrogen (N) and phosphorus (P), which are not significantly removed except in tertiary treatment. These elements in wastewater may increase the value of the water for irrigation, but they increase its pollution and eutrophication. Not all of the fertilizer value can be used, especially because irrigation late in the season often occurs at a period when consumption of nutrients by plants is low. The soil profile is thus enriched with nutrients subsequently leached by winter rain into the ground water.

Water quantity and water quality, again, are very closely related. Reductions in water quality can reduce the available supplies for almost any use just as drought does. In addition, it is almost always cheaper to prevent deterioration of water quality in the first place than to remediate the problem once deterioration has occurred. Maintaining water quality in effect augments available water supplies. In many instances, measures to maintain water quality are technically feasible. Economic feasibility varies with the situation and with the technology or management protocols employed. Environmental quality will almost always be maintained or improved through efforts to maintain water quality. Finally, active maintenance of water quality will almost always benefit both present and future generations.

Watershed Management

Watershed management is defined as the art and science of managing the land, vegetation, and water resources of a drainage basin to control the quality, quantity, and timing of water for enhancing and preserving human welfare and nature. By altering the natural hydrologic cycle, the physical diversion of water often has undesired environmental consequences. This effect can best be illustrated by looking at the consequences of existing regional large-scale water supply projects. The management of Lake Kinneret/Lake Tiberias/Sea of Galilee as a major water supply reservoir and the diversion of the Yarmouk River have almost eliminated the inflow of freshwater to the lower Jordan River. The river has become saline, with a resulting loss of freshwater flora and fauna. Similarly, development of ground-water supplies near the Azraq Oasis in eastern Jordan has eliminated the flow of the two freshwater springs feeding the oasis (Salameh and Bannayan, 1993), again with the resulting loss of freshwater flora and fauna. Drainage of the lake and marshes in the Hula Valley in northern Israel resulted in soil oxidation and subsidence, as well as the loss of native fauna and flora.

With increasing environmental awareness in the study area, the societal and environmental impact of watershed management, as indeed of all schemes to increase water supplies, has become an essential consideration in the planning process. The committee believes that a larger land-scape approach is essential for true watershed management.

Dams

Most watershed management schemes in the area involve small-scale efforts to capture stormwater runoff in dams, ponds, or retention basins. Stormwater runoff in the study area is intermittent and highly variable spatially as well as temporally. In rainy years the runoff can reach hundreds of million m^3/yr, while in dry years it can be negligible. Almost all of the wadis in the Jordan Valley that drain to the Dead Sea, all of the wadis in the Rift Valley south of the Dead Sea, and many of the wadis draining to the Jordan River are not dammed. Some of the wadis, particularly in the south, contain brackish ground-water discharge in their lower reaches, but in their middle and upper reaches all the wadis are dry except during and immediately following storms. Whenever possible, decisions concerning the management of the area's water resources should not be made on a small-scale, short-term, site-by-site basis, but instead to promote the long-term sustainability of all aquatic resources in the land-scape.

Large dams in the wadis, such as the King Talal Dam on the Zarka River, generally cannot be filled with water generated within their own watershed areas (Salameh and Bannayan, 1993, p. 22). They can only be managed efficiently if water is imported into the watershed, and thus will likely be limited to the northern, more developed parts of the study area. The quality of water stored in multisource reservoirs needs to be considered during the planning phase, because it may affect the usefulness of the stored water. The King Talal Reservoir contains stormwater runoff (some of which originates in the Amman area), springwater (some of which is saline), and both treated and untreated sewage effluent. The reservoir was originally planned for potable water for the Amman area, but because of the poor quality of the water, its use is now restricted to irrigation in the Jordan River Valley. Similarly, the Baruch Reservoir in the Yezreel Valley in Israel contains treated wastewater and saline spring discharge as well as stormwater. This reservoir and some small ponds in the valley have created a severe drainage and salinity problem (Binyamin et al., 1991). One option to increase the reliable water supply from the Yarmouk River is to construct the proposed Unity Dam. However, the committee did not analyze this option because it involves issues beyond the scope of its study.

Small Retention Structures

In contrast to large dams, small retention structures on the wadis could possibly be effective in capturing stormwater runoff. In Israel, the quantity of stormwater runoff that can be captured is estimated at 160 million m³/yr (Soffer, 1992). Presently, only 25 percent of this water is intercepted or stored underground, in approximately 120 small open reservoirs spread throughout the country. The total storage capacity of these reservoirs is about 100 million m³, sufficient to store most of the available runoff from the watersheds.

According to the Palestinian Water Authority, the quantity of stormwater runoff in the West Bank is 70 million m³/yr. Only 2 million m³/yr of stormwater runoff is generated within the Gaza Strip, although floodwaters originating in Israel make the total available runoff about 15 million m³/yr. None of this water is presently intercepted, but several feasibility studies are being considered. An estimated 13 to 15 million m³/yr (BRL-ANTEA, 1995) could be captured through construction of storage structures on four of the principal wadis draining eastward in the West Bank (Wadis Fara'a, Badan, Maleh, and Qilt). Studies have also begun on recharging captured flows in Wadi Gaza into the coastal aquifer.

About 40 million m³/yr of stormwater runoff may be available from wadis in Jordan that are tributary to the Jordan-Dead Sea-Rift valley (Open files, Water Authority of Jordan). Pilot projects on four of the wadis are under way (Kifaya, 1991) to test the feasibility of constructing small retention structures.

The utility of water stored behind retention structures in many of area wadis will be limited by the structures' remoteness from urban and agricultural areas and the fact that they will likely hold water only for a short period of time. If the water they capture is not used shortly after storms, much if not all of the water will be lost to evaporation, transpiration, and seepage, and will therefore be unavailable during the dry season. One possibility is thus to use retained stormwater for artificial recharge. This strategy has proven practical for recharging unconfined aquifers in arid areas (U.S. Army Corps of Engineers, 1979). Several stormwater interception projects in Israel are used to recharge ground water (although none presently in the eastward-draining wadis). Among existing projects are the Shiqma Reservoir in the south and the Menashe project in the north. These projects utilize infiltration of retained stormwater to recharge shallow, unconfined aquifers. Stormwater might also be used for the irrigation of single trees or small fields in the arid southern part of the study area.

Although direct infiltration to unconfined aquifers may not be possible at many potential stormwater retention sites, recharge to underlying confined aquifers by means of injection wells may be a possibility. Where head relationships are suitable, recharge could be accomplished by gravity drainage, which would be a major advantage in remote areas. Particularly in the wadis draining toward the Rift Valley, the recharged water would most likely create a storage zone of fresher water (compared to ambient water quality in the aquifer), which could be withdrawn when needed. This process is called aquifer storage and recovery. Problems of sediment removal and clogging of wells would need to be studied.

Urban runoff is another source of water for retention basins. Impervious urban areas in arid environments can sporadically generate large quantities of runoff, which, depending on the degree of treatment, can be used for various types of irrigation or for ground-water recharge (Ishaq and Khararjian, 1988).

However, urban runoff, especially from industrial areas, may not be suitable for ground-water recharge and further treatment before recharge may be necessary. Current practices within the study area are to convey urban stormwater to natural drainage outlets or to collect it in the sanitary sewer system. Retention basins, in addition to storing usable water, would therefore attenuate flooding and avoid excess flows at wastewater treatment plants. Finding suitable locations in urban areas may be a problem, and retrofitting existing collection systems would be expensive. However, the expansion of existing urban areas and the possible creation of new urban centers to accommodate regional population growth afford the opportunity to plan for alternative and beneficial uses of urban stormwater runoff. Planning should involve consideration of the treatment required to make the water suitable for its intended use, and the downstream environmental effects of diverting storm runoff away from natural drainage ways.

Altering the Hydrologic Cycle

Methods for decreasing evapotranspiration, increasing natural ground-water recharge, and decreasing natural ground-water discharge, although less common than the capture of stormwater runoff, are other possible watershed management options. Evapotranspiration rates are highest in wetlands and over open water bodies. The drainage of the Hula Wetlands in northern Israel to create arable land and the near elimination of the Azraq Oasis by ground-water pumping are examples of unintentional decreases in evapotranspiration, with equally unintentional environmental consequences. (See Boxes 4.1 and 4.2.) Thus, the elimination or reduction of wetlands and open water bodies, although immediately attractive to meet water demand, will almost always have a negative effect on biodiversity and the natural functioning of the ecological system and should therefore be approached with caution.

Natural recharge can be increased by dewatering aquifers that have natural water tables within one or two meters of land surface. Not only will this approach decrease evapotranspiration, but potential recharge formerly rejected by the aquifer when the ground or soil was saturated will now be able to infiltrate into the aquifer. Such planned management schemes are related to and have similar environmental consequences as other schemes to decrease evapotranspiration.

Pumping of ground water results in a reduction of natural discharge and a decrease in water levels (pressure) within the aquifer. In aquifers adjacent to brackish or saline water, the decline in pressure results in the movement of this poorer quality water into the freshwater part of the aquifer. In the study area, lateral movement of seawater has occurred in the coastal aquifers in Israel and the Gaza Strip (Palestinian Water Authority, MOPIC, 1996), and upward migration of saline water has occurred in the aquifers underlying the Dhuleil and Badia areas in Jordan (Salameh and Bannayan, 1993). Projects to intentionally decrease groundwater discharge should be carefully evaluated on the basis of predictable hydrologic consequences. However, on a small scale and with suitable physical conditions, ground-water discharge may be decreased and water levels increased by the construction of underground dams (Nilsson, 1988). Mediterranean coastal areas underlain by thin, unconfined sand aquifers might be particularly suited to this technology, which generally involves the construction, by injection through closely spaced boreholes, of a cement or low-permeability grout curtain extending to the base of the aquifer. In addition to storing ground water, underground dams can also prevent the lateral intrusion of saline water into coastal aquifers (Garagunis, 1981). Another method is keeping a water table mound along the coast while lowering the water table by pumping further away from the coast, a method successfully used in the coastal aquifer of Israel.

The attractiveness of watershed management will depend critically on the local circumstances and the management measures proposed. Table 5.5 summarizes the potential of watershed management measures in view of the five committee criteria. In most instances, watershed management will increase available water supplies. Situations in which management measures are undertaken to improve water quality may be an exception. There are a whole array of technically feasible watershed management techniques, whose economic feasibility will vary from situation to situation, depending in part on the value of the additional increments of water produced. The environmental impacts of watershed management will also be situation-specific. In general, where watersheds are managed to improve or maintain water quality, the environmental im-

TABLE 5.5 Augmenting Supplies

Committee Criterion	Watershed Management	Water Harvesting	Ground-Water Overdraft
1. Impact on Available Water Supply	+/–	+	+ (short term)
			– (long term)
2. Technnically Feasible	+	+/–	+
3. Environmental Impact	+/–	+/–	-
4. Economically Feasible	+/–	+/–	+ (short term)
			– (long term)
5. Implications for Intergenerational Equity	+	?	–

NOTE: + indicates positive effects, – indicates negative effects.

pact will be salutary. Where watershed management involves the construction of large dams, environmental impacts may be negative. Finally, inasmuch as watershed management activities enhance the quantity or the quality of water available, or both, the impact on present and future generations will likely be positive.

Water Harvesting

Rainfall is the ultimate source of all freshwater in the study area. The direct capture of rainfall is referred to as water harvesting. The most common methods of water harvesting are the use of rooftop cisterns for individual domestic supplies, and catchment systems and storage ponds for agricultural supplies. Cisterns are used throughout the world for rural water supplies, and they are particularly common in villages within the West Bank (Anonymous, 1988). Investigations by the Palestinian Hydrology Group (PHG, 1992) and Bargouthi and Deibes (1993) indicate that 45 percent of the rural areas (37 percent of the population) in the West Bank depend on rainwater harvesting to satisfy their basic water needs. The increased use of cisterns for domestic supply has been suggested for the Gaza Strip (Abu-Safieh, 1991) and Jordan (Tekeli and Mahmood, 1987).

Cisterns used in the West Bank and Gaza Strip are generally small excavations in the ground, generally no more than 6 m in depth. They are of various shapes but the traditional shape is like an inverted cone, or *najasa*, meaning pear-shaped. A typical cistern can store 70 to 100 m^3 of rainfall, which is accumulated from rooftop catchments during the winter. This quantity is sufficient to satisfy the needs of a family of five for the dry summer months. Moreover, the cost of this water is almost zero, if the initial cost of construction is not considered. Even with construction costs, however, the cost of cistern water will remain cheaper than water purchased from the conventional water distribution system or from water tanks in areas not served by distribution systems. For example, the Palestinian Hydrology Group (PHG, 1995) has reported that a cistern can save an average family 1,000 to 1,500 NIS annually (about 12.5 percent of its annual income). Even where conventional sources of water are available, cisterns can provide supplemental water inexpensively and relieve the demand on the water distribution system.

In rainfall catchment systems, quality problems may be associated with the first water to enter the cistern ("the first flush") following the onset of rainfall (Yaziz et al., 1989). However, various methods, such as first-flush diversion devices, roof maintenance, removal of overhanging vegetation, and installation of screens, can be used to protect and improve the quality of the stored water (Krishna, 1991). Other technologies are also available to provide point-of-use (at the tap) or point-of-entry (at the house) treatment (Rozelle, 1987), as described later in this chapter.

The expansion of existing urban areas and possible creation of new urban centers in the study area may allow rainfall collection systems to be incorporated into housing designs. Large-scale urban construction projects can incorporate systems to convey water from rooftop catchments to centralized treatment plants for inclusion in municipal water supplies.

The use of ponds to store rainfall for livestock and irrigation is common in the Gaza Strip. These ponds are generally cubic or trapezoidal, with volumes of as much as 300 m^3 for cement ponds and as much as 3,000 m^3 for earthen or plastic ponds. Use of the ponds has partially replaced pumping from the shallow aquifer in the Gaza Strip and has helped prevent further deterioration of ground-water quality. In some cases, brackish ground water is mixed with collected rainwater to produce water that is suitable for irrigation. The use of these ponds as a source of water for artificial recharge in the Gaza Strip has been suggested by Al-Khodari (1991). Agricultural and artificial recharge designs that incorporate both water harvesting and the capture of stormwater runoff may also prove feasible in the study area.

The promise of water-harvesting options is very situation-specific, varying with both location and the particular activity proposed. As indicated in Table 5.5 water harvesting by its very nature will augment the available water supply. However, technical and economic feasibility and likely environmental impacts cannot be readily judged without more detailed information about the particular water-harvesting measure and the place it will be used. It is difficult to assess the implications of this approach for future generations in the absence of specific details.

In the 1950s, it was demonstrated in the Negev Desert that a system of capturing low natural rainfall without irrigation was technically feasible for growing food crops (Amiran, 1965; Evanari et al., 1982). This was a system developed by Nabatean and later Byzantine farmers in the first millennium A.D. that later fell into disrepair, and is no longer regarded as economic. It involves arrangements of cisterns, channels, soil, rocks, and plants to maximize the amount of rainwater available to plants during their growing season.

Ground-Water Overdraft

Because of the temporal variability of ground-water recharge, extraction of ground water by pumping almost always results in at least a seasonal decrease in the resources quantity, but this quantity will fluctuate around the long-term average and, if extraction does not exceed recharge, the supply is sustainable. If, on the other hand, extraction rates continually exceed recharge rates, ground-water levels will decline and the aquifer is overexploited. The amount of extracted water that exceeds the recharge rate represents a nonrenewable resource.

Ground-water overdraft is significant in Jordan, for example the Azraq Basin (Box 4.1), and efforts should be made to avoid further depletion of the resource. Future water supply scenarios in the area (CES Consulting Engineers and GTZ [Association for Technical Cooperation], 1996) indicate an overall reduction in conventional (renewable) ground-water withdrawals west of the Jordan River (Table 5.6). Ground-water mining of an aquifer that is hydraulically connected to a saline water body, such as saline ground water or seawater, will result not only in depleting the resource, but in degrading the quality of the freshwater. Overexploitation of the coastal aquifer in Israel and the Gaza Strip, for example, has resulted in the landward encroachment of saline water. Because of the almost immediate environmental consequences of ground-water overdraft and the eventual environmental and water quality consequences of the total depletion of the resource, attempting to reduce extraction rates from overexploited aquifers should be a high priority in the study area.

With proper management, most aquifers can provide a sustainable water supply and the basis for maintaining ecosystem biodiversity, as long as they are recharged. To ensure the sustainability of this ground-water use, research is needed on the amount of water held in storage and the environmental consequences of depleting this storage. More consideration also needs to be given to beneficially using the storage space created by ground-water mining. This space can be utilized to store freshwater from a variety of sources through artificial recharge. Freshwater from remote sources and treated wastewater have been used to artificially recharge the coastal aquifer in Israel. In addition to being stored in

TABLE 5.6 Consolidated Conventional Water Resources

Sub-region	Current Surface Water Developed (million m³/yr)	Future Surface Water Developed (million m³/yr)	Current Ground Water Developed (million m³/yr)	Future Ground Water Developed (million m³/yr)	Current Total Water Resource (million m³/yr)	Total Water Resource (million m³/yr)
West of JRV	685	725	1,234	1,299	1,919	2,024
East of JRV	290	475	535	488	825	963
Total Study Area	975	1,200	1,769	1,787	2,744	2,987

NOTE: JRV, Jordan Rift Valley; Future, Year 2040.

SOURCE: Reprinted, with permission, from CES Consulting Engineers and GTZ, 1996. ©1996 by Consulting Engineers Salzgitter GmbH.

the aquifer for later use, this water has also been used to control water quality degradation in the aquifer. Similar schemes have been proposed for the coastal aquifer in the Gaza Strip and elsewhere (Assaf, 1994).

A special case of overexploitation occurs in aquifers that receive little or no recharge. Water contained in such aquifers is referred to as "fossil" water and is considered a totally nonrenewable resource. The Disi Aquifer in southern Jordan and the Nubian Sandstone Aquifer in southern Israel are the principal geologic units containing fossil water in the study area. These thick sandstone sequences were filled with freshwater during the geologic past, when climatic and geologic conditions were more favorable for recharge. They currently receive little or no recharge.

It is estimated that billions of m^3 of good to excellent quality water are stored in the Disi and Nubian aquifers. According to the Multilateral Working Group on Water Resources (CES Consulting Engineers and GTZ, 1996) 253 million m^3/yr of fossil ground water are projected to be available in the study area (110 million m^3/yr west of the Rift Valley and 143 million m^3/yr east of the valley). This rate of withdrawal would indicate a "life expectancy" of the resource measurable in hundreds of years. Current withdrawals are at the rate of 95 million m^3/yr (25 million m^3/yr west of the Rift Valley and 70 million m^3/yr east of the valley). However, the consequences of the current or higher withdrawal rates are largely unknown.

What is known is that as water is withdrawn from aquifers containing fossil water, a continual decline in water levels occurs. This decline induces saline or brackish water from adjacent rocks to move into the aquifers to replace the freshwater removed from storage. In addition freshwater from overlying aquifers may be induced to move into the fossil aquifer. Research is needed on how these phenomena will affect the sustainability of water resources in the study area. The extraction of fossil ground water may not have significant environmental consequences because the aquifers are isolated from the biosphere and have little or no natural recharge or discharge. However, because this resource represents the largest untapped water resource in the study area, it should be developed within the framework of an overall sustainable development scheme. Options for the development of fossil ground water include use for local development (the current practice), use as part of a national or regional distribution system, or retention as a reserve water supply. An economic analysis of the consequences of the development or deferred development of fossil ground water needs to be made. Such an analysis should be based on hydrologic research and should include consideration of the environment and the protection of future generations. As mentioned in Chapter 2, temporary overpumping in an aquifer can provide time for transition to a nonagricultural economy.

As shown in Table 5.5 ground-water overdraft is not an especially attractive means of augmenting supplies, except in some special circumstances. Ground-water overdraft is always self-terminating, as water levels ultimately decline to depths from which it is no longer economical to pump. Thus, ground water overdraft can augment available water supplies in the short run but if overdraft persists, the quantities of water available decline. Ground-water overdraft is technically feasible, except where the depth to the water table is very great. The environmental impacts of ground-water overdraft, such as subsidence and seawater intrusion, are almost always negative.

The economic feasibility of overdraft varies over the period of exploitation. In the short term, there may be temporary circumstances, such as the need to combat drought, when overdrafting may be economically feasible and economically optimal. However, overdrafting is not economically optimal over long periods of time, and ultimately it becomes economically infeasible when water tables are drawn down to levels from which it is not economical to pump. Overdrafting will almost always have negative impacts for future generations, since the stock of ground water left for future generations will always be smaller than if overdrafting had not occurred in the first place.

Wastewater Reclamation

Clearly, water is too precious a commodity within the study area to be used only once and then discarded as a waste product (similar U.S. areas are discussed in Boxes 5.2 and 5.3). Indeed, the study area already is a leader in the use of reclaimed wastewater for nonpotable use. About 53 percent of the total urban (domestic) water used in the area receives some form of treatment after use. In Israel, about 70 percent of all the effluent from municipal wastewater treatment plants is recycled (Argaman, 1989). In Jordan, about 60 percent is recycled (WAJ files). Almost none is being treated in the West Bank, and nearly 20 million m^3/yr is treated in the Gaza Strip. As the demand for water continues to increase beyond the available natural supply, it is not unreasonable for water supply plans to forecast a near-total reuse of water in the study area.

Reclamation has two benefits. The first is pollution abatement—the elimination of wastewater as an environmental and health hazard. The second is source substitution—an increase in the net amount of water available for use (U.S. EPA, 1992). For this reason, wastewater reclamation is more common in arid and semiarid dry regions and countries, such as Australia, the western United States, Mexico, the Arabian Peninsula, South Africa, India, Cyprus, Tunisia, and Israel.

**BOX 5.2 Harlingen, Texas: The Importance of
Source Water in Industrial Reuse**

The city of Harlingen, Texas, is located 16 km north of the U.S.-Mexican border
in the southern tip of Texas. In 1988, the city recognized that it needed additional
water supplies if it was to attract new industry to create additional employment and
increase tax revenues. The city undertook a study to identify alternative means of
increasing the water supply, and three were identified. First, it could purchase potable water from neighboring communities, but at a very high price. Second, it could
acquire low-quality irrigation water supplies from nearby growers. Third, it could
reuse treated municipal wastewater. Domestic wastewater reuse was found to be the
most cost-effective source of additional supply, despite the fact that existing treatment facilities would have to be modified and some new facilities constructed.

The city was fortunate in that existing wastewater streams could be segregated at
very low cost. Existing industrial wastewater was of very low quality, with prevailing
total dissolved solids levels so high that treatment with reverse osmosis was not
feasible. Domestic wastewater, on the other hand, was of sufficiently high quality
that reverse osmosis could be used to produce relatively high-quality source water
for industrial use. In this case, the key to the cost-effectiveness of the wastewater
reuse system lay in the fact that existing sewer and wastewater disposal systems did
not have to be retrofit in order to separate domestic and industrial wastewater
streams. Had retrofit been necessary, the plan would not have been cost-effective.

The plan ultimately adopted entailed conventional treatment of domestic wastewater, followed by coagulation and filtration to reduce suspended solids, and reverse
osmosis. Industrial waste was treated in a new and separate facility and discharged.
The new wastewater reuse system was placed in operation in 1992. This reuse
facility produced 7,400 m^3/day of demineralized water at a unit cost of $0.23/$m^3$.
The capital costs of the project totaled $9.5 million, while annual operating costs
amounted to $400,000. The experience at Harlingen illustrates the importance of
source water quality in determining whether wastewater reclamation and reuse
schemes can be feasible and cost-effective.

Reclamation will also serve to increase the usable amount of any
additional supplies obtained from other sources of freshwater. For example, 50 million m^3 of fossil water used for municipal supply can result
in an additional 35 million m^3 reclaimed water (assuming a 30 percent
reduction due to conveyance losses and consumptive use).

There are three categories of wastewater reclamation: (1) direct use of
wastewater with little or no treatment; (2) direct use of wastewater after
suitable treatment; and (3) the indirect use of treated effluent after suitable treatment.

Direct Use of Untreated Wastewater

Some types of untreated wastewater, such as wash water from washing machines and kitchen sinks (so-called "gray water") can be used di-

BOX 5.3 Multiple Use of Reclaimed Wastewater in Southern California

The West Basin Municipal Water District wholesales water supplies for some 900,000 residents and a large number of industries on the western side of the Los Angeles Basin. Throughout its history, the district has been confronted with an acute scarcity of local supplies and high cost and scarcity of imported supplies. With several other water purveyors in southern California, the West Basin District has been an innovator in utilizing reclaimed wastewater to meet the demands of its customers. In recent years, the incentive to employ wastewater reclamation has been strengthened, as coastal communities in southern California have come under intensifying pressure to reduce the quantity of wastewater discharged to the Pacific Ocean. This pressure has worked to the advantage of the West Basin Municipal Water District, because its service area lies in proximity to the Hyperion Wastewater Treatment Plant, the largest treatment plant of the City of Los Angeles.

The West Basin District has embarked on a decade-long project that will ultimately reclaim 86 million m^3 of wastewater annually. The first phase of this project, to be completed shortly, will yield 20.5 million m^3 annually. Initially, product water will be distributed evenly among three different uses. Three treatment streams have been designed to match the quality of the product water for each of the uses. The first, and ultimately largest, use is for seawater intrusion control. These waters are injected into the West Basin Aquifer to form a freshwater barrier between inland ground water and intruding seawater. The barrier injection water will undergo biological denitrification, softening, pH adjustment, filtration, reverse osmosis, and disinfection. The product water will meet all state standards for drinking water.

Second, a portion of the reclaimed water will be used for industrial purposes. Two large oil refineries will use product water as a cooling water makeup source, and for boiler feedwater and other process uses. This water will be denitrified, softened, filtered, and disinfected, and will meet standards for industrial use in petroleum refining. Third, product water will also be used for public landscape irrigation. Over 1,200 users have been identified, including public parks, schools, cemeteries, and golf courses. Treatment for this use consists of coagulation, flocculation, filtration, and disinfection. This product water will meet all state requirements for unrestricted irrigation reuse of wastewater. The unit cost of the water averaged across all uses is $0.57 m^3, which is considerably less expensive than alternative sources of supply.

The West Basin Water Reclamation Program illustrates how a large wastewater reclamation project can be tailored to serve a number of uses, each requiring a different level of water quality. In general, such multiple-purpose projects will be most economical where large quantities of wastewater are available for reclamation, allowing economies of scale to be realized.

rectly in homes for toilet flushing or garden irrigation. Gray water can contain contaminants that are harmful to humans, and therefore care is needed in separating gray water plumbing from potable water plumbing.

Direct use of untreated wastewater is possible in some industries by in-plant industrial recycling; its adoption can be encouraged by pricing or permitting policies for water supplied to industries and by imposing quality standards and impact fees on industrial effluent. Water quality requirements of individual industrial applications will determine the practicality of direct reuse. Another possibility for direct reuse is the reapplication of agricultural drainage (return flow) for irrigation. For agricultural applications, the quantity of directly reused water will be limited by water quality considerations, although blending or cycling drainage water with better quality water can increase its usefulness (Grattan and Rhoads, 1990).

Direct Use of Treated Wastewater

Although it is technologically possible, treated wastewater is not used directly for potable supplies, because the public generally does not accept the concept (U.S. EPA, 1992, p.106). Nevertheless, after suitable treatment, wastewater is used directly for many nonpotable uses in all three major sectors: urban, agricultural, and industrial. Table 5.7 shows the physical and chemical characteristics of the effluent from the Dan District treatment facility (Shafdan) in Israel following infiltration and pumping

TABLE 5.7 Quality of Effluent Following Advanced Treatment (Soil Aquifer Treatment) the Shafdan Project, Israel[a]

Parameter	Unit	Post-Treatment Value
Total dissolved solids	mg/l	1031
Cl	mg/l	322
Sodium	mg/l	227
EC	ds/m	1.76
SAR	$(meg/l)^{1/2}$	5.1
Boron	mg/l	0.5
pH	units	7.73
NO_3-N	mg/l	5.3
NO_2-N	mg/l	2.9
Total N	mg/l	8.8
Alkalinity (as $CaCO_3$)	mg/l	310

[a]All other parameters—biological oxygen demand, coliform bacteria, viruses, trace elements, trace organics, and toxic substances—were very low.

SOURCE: Reprinted, with permission, from Kanarek et al., 1994. ©1994 by Mekoroth Water Co. Ltd.

from a shallow aquifer (Soil-Aquifer Treatment). The water is of better quality than the regular drinking-water supply.

Urban. Although reclaimed wastewater will probably not be used as a potable source, it has many other potential urban uses. Urban nonpotable uses include landscape irrigation, toilet flushing, construction, vehicle and street cleaning, fire protection, and air conditioning (Okun, 1994). Urban reuse of wastewater requires dual municipal distribution systems—one for potable water and the other for reclaimed wastewater. Although major retrofitting of existing urban infrastructures would be costly, its cost must be compared to the cost of providing additional potable water from alternative sources. The prospect of major urban expansion in the study area may allow the planning of communities with an initial dual water system. Variability of supply and demand may require storage facilities, and the need for a noninterruptible supply may also require multiple treatment plants (U.S. EPA, 1992). Because urban and industrial effluent is enriched in dissolved solids, repeated wastewater recycling will significantly increase salinity. Although membrane and other expensive treatment processes can remove salts, efforts to minimize the salinization of wastewater are a requirement for sustainable urban recycling.

Agriculture. By far the largest use of reclaimed water is that of the agricultural sector. In Israel in 1994, 254 million m^3/yr of reclaimed water was used to irrigate more than 27,000 ha (Table 5.8) (Eitan, 1995); see Box 5.1 for further discussion. This source therefore represents about 65 percent of total agricultural water use. The Israeli Water Commission esti-

TABLE 5.8 Crop Areas Irrigated With Treated Effluent in Israel (ha), 1994

District	Crops	Field Citrus	Crops	Tree Crops	Forage Various	Total
Jerusalem	3,103	0	72	0	0	3,175
North	5,653	106	282	975	92	7,078
Haifa	4,568	121	38	45	0	4,772
Center	4,752	394	65	1,188	102	6,504
Tel Aviv	307	48	17	0	0	372
South	4,157	25	126	1,096	280	5,684
West Bank	16	0	0	0	0	16
Gaza	7	0	6	0	0	13
Total	22,533	694	606	3,304	474	27,611

SOURCE: Eitan, 1995.

mates that, by the year 2020, 782 million m^3/yr of treated effluent will be produced in the area west of the Jordan River and that 98 percent of this effluent (767 million m^3/yr) will be used for irrigation. In Jordan in 1994, 59 million m^3/yr of reclaimed wastewater was used in irrigation, representing 8 percent of the total agricultural use (see Table 2.3); however, this percentage is expected to increase because Jordanian legislation, strengthened in 1995, requires that all new sewage treatment plants include a provision for reuse of the effluent, with emphasis on expanding agriculture in the eastern uplands (WAJ open files). Irrigation with reclaimed water has been suggested as well for the Gaza Strip and West Bank (Abu-Safieh, 1991; Sbeih, 1994), but awaits improvement of wastewater treatment facilities. Currently about 20 million m^3/yr of wastewater are treated in the West Bank and Gaza Strip, but the quantity is expected to increase to about 43 million m^3/yr with sewerage expansion and increased water consumption. Substituting reclaimed for potable water in irrigation obviously allows the more beneficial and efficient use of the limited freshwater sources.

Forage and other crops that are not consumed by humans have the lowest water quality requirements, and for irrigation of these crops, effluent given only primary treatment is sometimes used. Primary treatment includes screening of coarse solids and grit removal, sedimentation of settleable material, and skimming of floatable material. But for most purposes, secondary treatment is required as well. Secondary treatment includes the use of stabilization ponds or aerated lagoons (low-rate processes) or activated sludge, trickling filters, or rotating biological contractors (high-rate processes). Generally, water after disinfection must follow treatment. Water after secondary treatment may be used for crops not directly consumed by humans; in Jordan, however, the quality of wastewater given even secondary treatment may preclude the use of this wastewater effluent on crops. For wide spectrum irrigation, tertiary (advanced) treatment is also required. Advanced treatment removes nitrogen, phosphorus, suspended solids, dissolved organic substances, and metals.

Permissible levels of suspended solids, biological oxygen demand, coliform count, and residual chloride for the irrigation of various categories of crops in Israel are presented in Table 5.9. Obviously, unrestricted irrigation requires higher degree of purification (category D) than irrigation of crops which are utilized only after processing (category A).

Aside from water quality considerations, the major problem in using wastewater for irrigation is the timing of supply and demand. Wastewater is a relatively constant source, whereas irrigation demand is variable. Thus, alternate uses and storage facilities for the treated water must often be included in the design of agricultural reuse projects. Storage facilities also provide for additional treatment of the wastewater. A properly man-

TABLE 5.9 Quality Criteria for Treated Wastewater Effluent to be Reused for Agricultural Irrigation in Israel[a]

	Crop Type			
	A	B	C[b]	D
Effluent Quality	(cotton, sugar beet, cereals, dry fodder, seeds, forest irrigation)	(green fodder, olives, peanuts, citrus, bananas, almonds, nuts, etc.)	(deciduous fruits, conserved vegetables, greenbelts, football fields, golf courses)	(unrestricted crops, including vegetables eaten uncooked (raw), parks, and lawns)
BOD_5,[c] total (mg/l)	60	45	35	15
BOD_5, dissolved (mg/l)	—	—	20	10
Suspended solids (mg/l)	50	40	30	15
Dissolved oxygen (mg/l)	0.5	0.5	0.5	0.5
Coliforms counts (100 ml)	—	—	250	12 (2.2)[d]
Residual available chlorine (mg/l)	—	—	0.15	0.5

[a]Requirements should be met in at least 80 percent of samples taken.
[b]Irrigation must stop 2 weeks before picking; no fruit to be taken from the ground.
[c]BOD measured or calculated over a 5-day period.
[d]Requirement should be met in at least 50 percent of samples.

SOURCE: Shelef, 1991.

aged sequential set of reservoirs can appreciably improve the quality of wastewater (Juanico, 1996). Storage of effluent in aquifers can provide the equivalent of tertiary treatment (Table 5.7; NRC, 1994).

Industry. Industrial uses of water include cooling, boiler-feed, and process water. All three of these uses can take advantage of reclaimed wastewater, although additional treatment, ranging from pH adjustment to carbon-adsorption filtration, may be required. Cooling water generally requires the least treatment and offers the most promise for expanded use of wastewater. Economics and public policy, however, may favor use of reclaimed water in the agricultural and urban sectors.

Indirect Use

Indirect uses of reclaimed wastewater include enhancement of the natural environment, fish farming, and ground-water recharge (U.S. EPA, 1992). Environmental enhancement, such as creation or augmentation of wetlands or lakes, and stream augmentation, requires water treatment commensurate with the degree of human contact with the water. Where sufficient treatment is provided, environmental uses of reclaimed wastewater may provide an alternative method of storing reclaimed water.

Commercial fish production in artificial impoundments of reclaimed water is widely practiced in Israel (Crook, 1990). Where fish are used for human consumption, the quality of the reclaimed water must be of sufficient quality to preclude the bioaccumulation of toxic contaminants (U.S. EPA, 1992).

By far the most significant indirect use of reclaimed water in the study area is for artificial recharge of ground water. Recharge may be accomplished by surface infiltration from impoundments or by injection from wells (NRC, 1994). Currently, only the infiltration method is practiced in the study area. The purposes for recharging ground water with reclaimed water include (1) providing storage for excess reclaimed water, (2) providing additional treatment, (3) replenishing aquifers, and (4) establishing saltwater intrusion barriers in coastal aquifers (U.S. EPA, 1992). These purposes are often synergistic. For example, secondarily treated wastewater from Tel Aviv is used to recharge an unconfined aquifer. Part of the recharge is used to replenish the aquifer, and part is later withdrawn and used for irrigation. Because of the additional treatment the water received within the soil zone and the aquifer, the withdrawn water is suitable for unrestricted irrigation of vegetables that are eaten raw, see Table 5.9 (NRC, 1994).

Ground-water recharge with surface infiltration systems is not feasible where poorly permeable soils are present, land is too costly, unsaturated zones have impermeable layers or contain undesirable chemicals that can leach out, or aquifers have poor quality water at the top or are confined (NRC, 1994). In these cases, ground water may be recharged by means of injection wells. Reclaimed wastewater can be used to replenish depleted, confined aquifers, to store water in brackish or saline aquifers for later withdrawal (referred to as aquifer storage and recovery), or create saltwater intrusion barriers. Repeated recycling of wastewater by means of surface infiltration or injection systems can lead to increases in ground-water salinity.

A closely related problem here and in connection with the overdraft of ground water is that aquifers in such areas as between Beersheva and the Gaza Strip are being considered for disposal of highly hazardous waste. Deep disposal could cause serious contamination of remaining supplies.

Cost of Reuse

There is sufficient technology and opportunity to allow for the total utilization of reclaimed wastewater in the area for the foreseeable future. The practical feasibility of a particular wastewater reclamation project is determined by its cost, which is dependent on the quantity of water involved, the quality of the source wastewater, and the desired quality of the treated water. Table 5.10 summarizes the characteristics of a number of wastewater reclamation projects in the United States and one in the region under study. For projects where very little treatment is required, such as Phoenix, Arizona, and Whittier and San Clemente, California in the United States, and the Dan Region Project in Israel, the costs are extremely modest. Source water for all of these projects is of relatively high quality and the product waters are used exclusively for ground-water recharge. The projects with relatively high costs, including Water Factory 21 in Southern California and El Paso, Texas, have relatively high treatment requirements. Product water from Water Factory 21 must meet standards for potable use, since drinking water is extracted from the re-charged aquifer. In El Paso, raw wastewater must be subjected to primary and secondary treatment prior to the advanced treatment to bring water quality to levels suitable for irrigation, ground-water recharge and industrial uses.

Thus, the costs of wastewater reuse are influenced by public health and environmental standards, which determine how much raw water must be treated prior to discharge. Where standards are high, the incremental cost of the treatment to make the water suitable for nonpotable uses may be quite reasonable (NRC, 1994). Where existing sources of surface and ground water are fully allocated, reclaimed water may also be economically attractive compared to alternative sources, regardless of prevailing wastewater treatment standards. Reclaimed water's economic attractiveness will of course vary depending on the costs of its required treatment and the costs of the alternatives. For this reason, the economics of developing reclaimed wastewater must be assessed case by case.

Table 5.11 summarizes the evaluation of wastewater reclamation against the committee's five criteria. It suggests that wastewater reclamation may be a relatively attractive option for the study area. Like recycling, wastewater reclamation can significantly add to the region's available water supply. Reclamation has proven technologically feasible not

TABLE 5.10 Costs of Wastewater Reuse[a]

Project Location	Date On Line	Reuse Volume (m^3/day)
Phoenix, AZ	1970s	56,800
Whittier, CA	1962	170,000
Dan Region Project, Israel	1977	33,000
San Clemente, CA	1970s	7,600
Harlingen, TX	1990	7,400
Chevron Refinery, CA	1996	18,600
Saudi Arabia	Proposed	292,000
Franklin Canyon, CA	1996	1,600
West Basin, CA	1994	236,000
El Paso, TX	1985	37,900
Water Factory 21, CA	1977	56,800

[a]Unit costs are in 1996 dollars.

[b]RW, raw wastewater; 2nd, secondary effluent; 3rd, tertiary effluent.

[c]1st, primary treatment; C, chlorination; CP, chemical precipitation; D, disinfection; GAC, granular activated carbon; F, filtration; N, nitrification; OP, oxidation ponds; RO, reverse osmosis; SAT, soil aquifer treatment.

only within the region but in many other places throughout the world. Its environmental impacts are generally positive because it improves water quality in addition to stretching water supplies through reuse. The economic feasibility of reclamation varies with such factors as the quality of the feedwater and the technology used. Reclamation is economically feasible in many circumstances, however, most likely including significant number of reclamation opportunities in the study region. Finally, by maintaining and enhancing water quality and allowing water to be recycled and reused, reclamation is a sustainable practice that will tend to enhance both the quantity and quality of water for present and future generations.

Use of Water of Marginal Quality

An unknown, but possibly significant, savings in freshwater could be obtained by substituting water of marginal quality in some activities that now use potable freshwater. "Marginal" is, of course, a relative term; reclaimed wastewater, for example, is a special case of marginal quality water. Brackish or saline ground water, in some cases even seawater, may be used for some of the just-described uses of wastewater. The discussion here is limited to brackish water, defined as having a chloride

Source Water Quality[b]	Project Treatment[c]	Type of Reuse[d]	Unit Cost ($/m³)
2nd	None	GW	0.01
3rd	None	GW	0.02
RW	OP, 2nd, SAT	GW	0.03[e]
2nd	F	GW	0.05
2nd	F, RO, D	IN	0.27
2nd	CP, F, D	IN	0.44
2nd	Variable	IR, IN, GW	0.18-0.42
2nd	CD	IR	0.46
3rd	N, F, RO, D	IR, IN, GW	0.57
RW	1st, CP, F, D, GAC	IR, IN, GW	0.76[f]
2nd	F, GAC, RO	GW	0.88

[d]IR, irrigation; IN, industrial; GW, ground water recharge.
[e]Operating costs only.
[f]Capital cost only.

SOURCES: Compiled from Asano, 1985; NRC, 1994; East Bay Municipal Utilities District, 1992; Filteau et al., 1995; Grebbien, 1991.

content greater than 400 mg/l or an electrical conductivity greater than 1.5 dS/m, and to the special case of potentially contaminated but otherwise potable water delivered to consumers.

The most common use of marginal quality water is of brackish water to irrigate crops that have a high tolerance for salinity. Much of the brackish ground water in the study area can be used directly for irrigation without desalination. The yields of some crops, such as strawberries or subtropical and deciduous orchards, are reduced when irrigated with water with an electrical conductivity greater than 1.5 dS/m, although the fruit may be of higher quality because of increased sugar content. Other crops, such as cotton and barley, are not affected at levels of 8 dS/m or more (Shalhevet, 1994). Management practices for using brackish water in agriculture include restricting the use of brackish water to tolerant crops and tolerant varieties, although the latter are not widely available; mixing water sources, when required, to achieve lower salinity; intermittent leaching; use of drip-irrigation technology when practicable; use of poor quality water only toward the end of the growing season; and avoiding irrigation during hot weather (Shalhevet, 1994).

Although the use of brackish water for irrigation can free up freshwater resources for other uses, this practice is not without drawbacks even with sound management practices. Assuming an irrigation efficiency of

TABLE 5.11   Reclamation, Marginal Water, and Desalination

Committee Criterion	Wastewater Reclamation	Use of Water of Marginal Quality	Desalination of Brackish Water	Desalination of Seawater
1. Impact on Available Water Supply	+	+	+	+
2. Technically Feasible	+	+	+	+
3. Environmental Impact	+	+/−	+	+/−
4. Economically Feasible	+/−	+/−	+	−
5. Implications for Intergenerational Equity	+	+/−	+	?

NOTE: + indicates positive effects, and − indicates negative effects.

75 to 80 percent, application of water with an electrical conductivity of 2.9 dS/m will result in salt accumulation in the soil to a level of approximately 4.4 dS/m in the soil saturation extract (Ayers and Westcot, 1976). Fortunately, there are a number of crops that can tolerate this level of soil salinity, among them cotton, barley, wheat, sorghum, beet, cowpea, zucchini, safflower, soybean, date palm, and many grasses. Where winter rainfall exceeds 400 mm/yr, salts accumulated during the irrigation season will be partially leached out, and salt accumulation in the soil during the following growing season will be about 3 dS/m. Under these conditions, some additional crops, such as broccoli, tomato, asparagus, and peanuts, may be irrigated directly with little likelihood of yield reduction. Leaching of salts from the soil zone, of course, results in an increase in the salinity of the underlying ground water. Use of brackish water (as well as wastewater) to irrigate fields overlying water table aquifers (such as the coastal plain and the Jordan Valley) is therefore not sustainable unless the salinity buildup is managed.

A preliminary analysis of brackish ground water in Jordan with salinity suitable for irrigation of salt-tolerant crops indicates that 70 to 90 million m³/yr may be safely withdrawn from aquifers on the eastern shore of the Dead Sea (WAJ open files). More than 300 million m³/yr of brackish water are available west of the Jordan valley (Goldberg, 1992); about 65 million m³/yr discharge to springs on the west shore of the Dead Sea within the West Bank (primarily the Feshka and Turieba Springs), about 40 million m³/yr are available in the Gaza Strip, and an additional 200 million m³/yr of brackish water is available throughout Israel.

With time, the quantity of marginal water in the area will increase because of saline water encroachment and the infiltration of pesticides, fertilizers, and wastewater into freshwater aquifers. For example, the chloride content of the coastal aquifer in Israel and the Gaza Strip is increasing at the rate of about 2 mg/l of chloride per year. Eventually, the problem of water supply from the coastal aquifer will be one of quality rather than quantity. The long term use of this aquifer and others with deteriorating quality will depend on finding suitable uses for marginal quality water.

Brackish water does not always exist where it could be used beneficially. However, it should be regarded as a valuable resource and transported to areas of use. For example, there are three separate distribution systems in the Negev Desert: the National Water Carrier, transporting freshwater; the ShafDan System, transporting treated wastewater; and a brackish water line transporting water from the Negev Plateau. Clearly, an expanded role for brackish water (as well as wastewater) will require similar engineering schemes throughout the study area.

Water supplies contaminated by inorganic and, in some cases, organic pollutants represent a special case of water of marginal quality. Contaminated water results not only from pollution of surface- and ground-water sources, but also from contamination of potable water distribution systems. In some cities, pipe distribution systems are in need of repair. The leaky pipes admit contamination when broken water mains are repaired and when water shortages depressurize the system. Water storage in homes, or on the roofs of buildings, provides breeding opportunities for harmful bacteria. It may be impractical to solve these problems by investing substantial amounts of capital into reconstructing the water distribution systems. Another possibility is to provide a final treatment for potable water just prior to its use.

Several competing technologies are in use for point-of-use (at the tap) and point-of-entry (at the house) water treatment. These range from natural coagulents (seeds of *Moringa olifera* and *Strychnos potatorium*) (Gupta and Chaudri, 1992) to reverse osmosis membranes (Tobin, 1987). Practicality depends on the cost and reliability of the technology. An educational program is also a prerequisite, for the consumer, the manufacturer, and the local water utility.

The following specific technologies in use fall into four categories: adsorptive filters, reverse osmosis, ion exchange, and distillation (Rozelle, 1987):

• **Adsorptive filters.** Adsorptive filters usually rely on granulated activated carbon (GAC). They reduce common tastes and odors, some turbidity, residual chlorine, radon, and many organic contaminants. Efficiency depends on the design. This method is probably the most cost-effective for point-of-use treatment. The carbon surfaces do provide opportunities for bacterial growth under stagnant conditions, but a U.S. EPA study of 180 homes reportedly showed that the GAC filters had no significant gastrointestinal and dermatological effects (Calderon et al., 1987).

• **Reverse osmosis (RO).** Reverse osmosis is the "high tech" method for reducing dissolved inorganics. It can remove some organics, depending on the type of membrane. Such units include, first, a particulate filter, followed by a GAC filter, the RO module, a water reservoir containing a pressurized rubber bladder, a final GAC filter, and a special spigot to the sink. The units operate solely on water main pressures between 40 and 70 pounds per square inch gauge (psig) (276 and 483 kilopascals [kPa]) on nonbrackish water up to 2,000 mg/l of total dissolved solids (TDS), and deliver up to 1 gpm (0.63 l/s). Removal performance depends on the type of membrane used, most commonly cellulose acetate or polymide.

• **Ion Exchange.** Ion exchange has been used for many years to soften water. It also reduces barium and radium, nitrates, some arsenates, and uranium anion.

• **Distillation.** Distillation is an effective method for producing contaminant-free water. Used in some water-bottling operations, the process is energy-intensive.

Maintenance is essential for these units to function properly. Particulate filters must be replaced before clogging and adsorptive media must be replaced before becoming saturated with contaminant. Replacement cycles depend on the water-use rate, the type of contaminant, and its concentration in feedwater. RO membranes typically operate for one to four years without membrane fouling or deterioration. Ion exchange units must be regenerated or replaced periodically. Distillation stills must be cleaned to avoid scaling.

Point-of-entry treatment is a major industry, supplying potable water to millions of consumers who are in isolated areas, on farms, or in communities where wells have been contaminated. The method is technically sound and economically feasible for reducing organic and inorganic contaminants. Controlling and monitoring these devices is the key to protecting public health.

Some states in the United States have established regulations that require a public utility type of organization for point-of-entry devices (Burke and Stasko, 1987). New York State has passed enabling legislation to form water districts that carry out point-of-entry treatment device programs in cases of private well contamination. Guidelines have been developed to ensure that the devices are properly installed, operated, and maintained by water districts, once the districts are organized. This program is expected to ensure safe drinking water for approximately three million people at a reasonable cost.

These devices clearly have their place for isolated users, but they could be used more widely to improve the overall standard of drinking water in the study area. They are more than an alternative form of treatment—they are flexible potential components of a water supply system. Use of these devices should be considered in the context of overall water supply and water reuse systems. Specific factors to consider include the institutional structure that would provide the devices and their maintenance, the quality of the raw water source and need for pretreatment, opportunities for contamination while water is in transport to the consumer, and the nature of the decentralized industry that would stem from widespread manufacture, installation, and maintenance.

The evaluation of using waters of marginal quality is also summarized in Table 5.11. By tailoring water supplies of differing quality to appropriate uses, it may be possible to use water of impaired quality that would otherwise remain unused. In this way, the available water supply would be increased. In many instances the use of marginal quality water is technically feasible. The environmental impacts of such use depend on the type and location of use. The continued recycling of irrigation tailwaters, for example, will ultimately lead to severe soil salinization, or, where salinity is actively managed, the need to manage significant quantities of saline drainage waters. The use of water of marginal quality is likely to be feasible where the water and the use are matched, so that substantial treatment costs are avoided. Finally, the use of these waters will tend to promote sustainability, stretching existing water supplies further, thus preserving the stocks and qualities of water available now and for future generations.

Desalination of Brackish Water

According to the International Desalination Association (1996), by December 1995, there were approximately 11,066 desalting units in operation worldwide, with a total capacity of 20.3 million m^3 per day (Figures 5.3 and 5.4). Considerable data have been assembled on technologies that are pertinent to the study area (e.g., Box 5.4; Awerbuch, 1988; and Hoffman, 1994). Small plants are already in operation in Israel (Glueckstern, 1991) and the Gaza Strip, and are being studied in Jordan (Fatafta et al., 1992). These plants have been used to evaluate technologies and refine cost estimates for large-scale desalination plants. The decision to introduce large-scale desalination into the study area depends on economics and on the success of other programs to manage demand or augment supplies. As long as other means of increasing the net availabil-

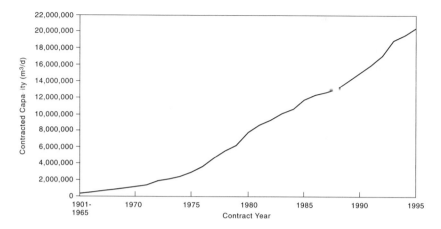

FIGURE 5.3 Cumulative capacity of all land-based desalting plants that can produce 100 m³ per day per unit or more of freshwater, by contract year. SOURCE: Reprinted, with permission, from Wangnick Consulting GmbH, 1996. ©1996 by Wangnick Consulting GmbH.

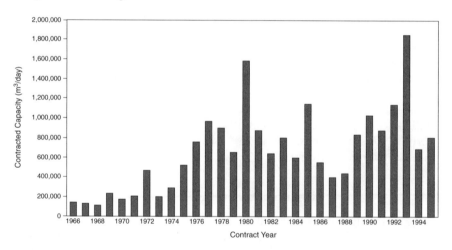

FIGURE 5.4 Capacity of all land-based desalting plants becoming active that can produce 100 m³ per day per unit or more of freshwater, by contract year. SOURCE: Reprinted, with permission, from Wangnick Consulting GmbH, 1996. ©1996 by Wangnick Consulting GmbH.

ity of water are less expensive, they will doubtless be developed to their fullest extent before desalination plays a major role in the area's water supply. According to Fisher et al. (1996) desalination will not be cost-effective in the study area until at least 2020.

One way to reduce the net cost of desalination is to tie the desalina-

---

BOX 5.4 Rural India: Brackish Water Desalination for Potable Use in Economically Developing Areas

Many arid and coastal regions of India must surmount severe water quality problems if reliable potable supplies are to be developed. These problems include high levels of salinity, nitrate concentrations, and pathogenic bacteria, in some cases, the presence of guinea worms. In the 1980s, a task force was assembled consisting of representatives from government, industry, and private organizations, to develop a potable water action plan. The resulting action plan was based on the following principles:

- A number of small plants, rather than a few centralized plants, would be required, since villages were highly dispersed over large areas.
- Standardized plant sizes, with capacities of 10, 20, 30, 50 or 100 m^3/day were to supply villages based on a designed demand of 10 l/day per person.
- Waters with total dissolved solids greater than 5,000 mg/l would undergo sand filtration, reverse osmosis, and disinfection.
- Waters with total dissolved solids less than 5,000 mg/l would undergo sand filtration, electrodialysis, and disinfection.
- Firms selected to design and construct plants would be required to (1) operate and maintain the plants for three years; (2) supply potable water to villages in the event of a plant shutdown; and (3) train local technicians to operate and maintain the plants.

Over 120 villages throughout India were selected to participate in the program. A total of 18 electrodialysis and 29 reverse osmosis plants were installed. Designs were kept as simple as possible due to the remoteness of many of the villages. Where feasible, local materials were used in construction. Standby diesel generators were installed at plants to avoid erratic power supplies in some villages, and emergency spare parts were stored at remote treatment sites. Where possible, plant brine was disposed of into nearby saline water bodies. Other disposal strategies included constructed evaporation ponds and ground-water injection.

This case illustrates how appropriately scaled technology can be employed to provide a potable water supply for communities and populations that are widely dispersed, where brackish water is readily available. The costs of the water vary with the quality of the source water and with the capacity of the treatment plant. In some instances, brine disposal may present problems, although evaporation ponds are generally effective in areas with high rates of evaporation. The success of this plan lies in appropriately scaling the technology to local conditions, use of local materials in construction, and employment of local residents to operate and maintain the plants.

---

tion process to other water projects. An example would be to substitute desalination for secondary and tertiary treatment of wastewater, so that the net cost of desalination would only be the cost above standard treatment. This approach might answer the public reluctance to accept reclaimed wastewater as a potable supply, and it may be practical if the availability of wastewater exceeds irrigation demand. Desalination of wasterwater would also help avoid salinity increases in soil and water. Offshore or coastal wells might also withdraw saline water for desalination. Where conditions are favorable, a line of saline water wells could create a pumping trough that would stabilize the movement of the freshwater-saltwater interface in a coastal aquifer. The protection of the freshwater aquifer would be an added benefit to the desalination process.

The lead time in evaluating brackish or saline sources of water, determining appropriate methods of brine disposal, and designing and constructing desalination plants must be considered in the overall planning. The environmental impacts of brackish or saline water extraction on the source water body and the impacts of disposal of the by-product brine on the receiving water can both be significant, particularly for large-scale projects.

Large supplies of brackish ground water exist throughout the study area, generally as part of complex ground-water flow systems, with freshwater occurring in shallow, or up-gradient, positions, and more saline water in deeper, or down-gradient, positions. The brackish water component of such flow systems cannot be developed without some long-term effect on adjacent fresh or saline components. Depending on the position of the wells in the flow system, withdrawal of brackish water results in either increased or decreased salinity. If salinity increases, the cost of desalination will generally increase; if salinity decreases, part of the adjacent freshwater resource will be depleted. Therefore, brackish water resources cannot be considered a "free good" or unlimited resource, but must be evaluated to determine their yield, salinity changes with time, and the effect of withdrawal on adjacent freshwater resources.

Another source of brackish water for desalination is agricultural return flow. Such flow constitutes the bulk of the water entering the Jordan River below Lake Kinneret/Lake Tiberias/Sea of Galilee. It has been suggested (Biswas et al., 1997, pp. 5.13-14) that drainage projects on both sides of the Jordan Valley could collect return flow that could be desalinated relatively inexpensively. Such a project would have the added benefit of rehabilitating Jordan Valley ecosystems if sufficient natural freshwater flows of desalinated water were maintained. Brackish ground water and agricultural return flows with salinities low enough for direct use as irrigation water should probably be reserved for this purpose.

Four distinct classes of technology are available for treating saline water. They are distillation technologies, electrodialysis, reverse osmosis, and salinity gradient solar ponds. With the exception of distillation, the cost of freshwater produced by each of these technologies depends on whether the raw water source is brackish water or seawater. The following sections describe those technologies most appropriate for treating brackish water.

Distillation. The oldest and best known desalination technologies are based on distillation. With these technologies, saltwater or brine is boiled, and the dissolved salts, which are not volatile, remain in solution as water is vaporized. When the evaporated water is cooled, pure water condenses. The Multi-Stage Flash (MSF) evaporation method (MSF) consists of a number of interconnected stages. Vapor pressure is maintained at each stage so that the boiling point of the feedwater is below its incoming temperature. Flashing occurs in steps as the feedwater passes in series through stages at successively lower pressures. The advantage of this system is that it can be operated at relatively low temperatures (70°C). A second distillation method, called the Multiple Effect (ME) evaporation system, uses evaporator chambers (the effects) that receive heat from an external source. The evaporation feed is preheated by hot product water from each effect. The process reuses heat very efficiently.

Distillation technologies are energy-intensive and the costs of producing product water are sensitive to energy costs. However, the costs do not vary with the quality of the feedwater and thus costs of desalting brackish water are approximately the same as for desalting seawater. Two technologies, electrodialysis and reverse osmosis, provide less costly means of desalting brackish waters.

Electrodialysis. The electrodialysis (ED) method of desalination is based on selective movement of ions in solution and the use of semipermeable membranes. When a current is applied, positive ions (cations) migrate to the negative pole, and negative ions (anions) migrate to the positive pole. A cation-permeable membrane allows cations to pass, but blocks anions. An anion-permeable membrane allows anions to pass, but blocks cations. Sets of the two types of membranes create alternate salt-depleted (product) and salt-enriched (brine) solution streams. Electrodialysis tends to be an attractive technology for desalting water with TDS concentrations of 3,000 mg/l or less, because its energy use in this range compares favorably with other technologies. Table 5.12 summarizes operating experience with electrodialysis units to desalt brackish waters in the United States and Canada. The data indicate that electrodialysis production costs are relatively high when compared to the costs of wastewater reclamation. Costs tend to be sensitive to the quality of the source water, although plant scale and specific type of technology used are also important.

Reverse Osmosis. The reverse osmosis process employs hydraulic pres-

TABLE 5.12 Costs of Brackish Water Desalination with Electrodialysis

Plant Location	Date On Line	Feed TDS (mg/l)	Capacity (m³/day)	Unit Cost ($/m³)
Yuma Proving Grounds AZ	1986	1,800	760	3.12
Granbury Water Treatment Plant TX	1985	1,625	490	2.33
Brazos Reservoir Authority TX	1989	321	10,980	1.24
Melville Water Treatment Plant Canada	1990	1,900	1,890	0.88
Foss Reservoir OK	1974	1,000	5,300	0.58
Robert House Reservoir VA	1990	7,570	690	0.50

NOTE: All costs in 1996 dollars. Total dissolved solids (TDS) of product water is less than 600 mg/l. Operating and maintenance costs normalized assuming an electricity cost of $0.04/kWh. Unit costs computed using a 20-year amortization period for capital costs and a 10 percent interest rate.

SOURCE: Reprinted, with permission, from National Water Supply Improvement Association, 1992. ©1992 by American Desalting Association.

sure to force pure water from saline feedwater through a semipermeable membrane (Box 5.5). The evidence suggests that reverse osmosis systems are most attractive for feedwaters with concentrations of total dissolved solids of between 3,000 and 40,000 mg/l. In this range, the reverse osmosis process uses somewhat less energy than electrodialysis. When total dissolved solids exceed 35,000 to 40,000 mg/l, multistage flash distillation becomes competitive in terms of energy use (Heitmann, 1990). Table 5.13 illustrates the costs and capacities of reverse osmosis facilities currently used to desalt brackish waters in the United States. In general, these costs vary with the scale of the plant and the technology used. The costs of facilities using relatively modern reverse osmosis technologies range from about $0.28 to $1.00 per cubic meter. The costs for desalting brackish water with reverse osmosis compare favorably with electrodialysis and with wastewater reclamation where reclamation requires extensive treatment. Generally, the costs of brackish water desalination also compare favorably with the costs of seawater desalination.

As summarized in Table 5.12, brackish water desalination may offer an attractive alternative to augment the water supplies of the region. Such desalination is technologically feasible and usually will not have adverse environmental impacts. Economic feasibility will depend on the quality of the feedwaters, the technology used, and the relative economic attractiveness of other alternatives. By stretching the existing water supply and providing qualitative improvements, brackish water desalination is likely to produce relatively favorable consequences for present and future generations.

Seawater Desalination

Seawater conversion tends to be very expensive, with costs ranging upward from approximately $1.00/m^3$. The costs of seawater desalination are significant, by setting an upper bound on the costs of additional water supplies for the Middle East and other areas with access to the sea. As far as we know, seawater desalination has the potential to provide virtually all the additional water needed in the study area, with possible negative environmental impacts limited to the effect of brine disposal on the receiving water. However, note that the seawater desalting costs reported here are for water at the desalting plant boundary. The costs of transport and pumping facilities must be added, along with the costs of operating and maintaining these facilities. Because desalting facilities will always be located at sea level, pumping costs can be especially significant, and seawater desalination will be most cost-effective in areas adjacent to the coast. The most cost-effective inland benefits of seawater desalination are likely to be the substitution of desalted water for freshwater supplies that will then be available for reallocation to other uses and other places. The comparative cost data suggest that, where wastewater is available for reclamation or significant supplies of brackish waters are available, it will generally be less costly to treat and reclaim these waters than to invest in seawater desalination.

Distillation

The traditional distillation technologies (discussed above) have costs that are extremely sensitive to energy prices, and, as a result, are only employed on any significant scale where there is no alternative source of water and energy is relatively plentiful.

Reverse Osmosis

In the last decade, reverse osmosis technology has been adapted for

BOX 5.5 Desalting Brackish Ground Water in Southern California

Continued population growth in southern California is placing intensified pressures on the region's water supplies. Nearly 75 percent of the water supplies available to southern California are imported. The prospects of developing additional sources of imported supply to serve population growth are uncertain. Competition for water to serve environmental purposes is intense, and the costs of additional imported supplies will be very high, probably $0.60/m^3$, and in many locales significantly higher. A number of local water purveyors in the southern California area have turned to brackish ground water as an alternative source of supply. The costs of reclaiming brackish ground water using reverse osmosis fall in the range of $0.28 to $0.45/m^3$, and thus compare quite favorably with the costs of new imported supplies. Three existing projects and one proposed project illustrate the variation in source water quality and uses.

Arlington Basin Desalting Project

The Arlington Ground Water Basin, approximately 80 km east of central Los Angeles, contains over 370 million m^3 of ground water that has been degraded with the residues of agricultural chemicals applied in the area over the past 70 years. Although the level of total dissolved solids (TDS) is only 1,100 mg/l, concentrations of nitrate-nitrogen exceed the existing standard, and dissolved organic residues of pesticides and herbicides are also present. In 1990, a pump-and-treat system employing reverse osmosis and activated carbon was placed in operation. The project produces potable water at a rate of 22,700 m^3/day. The water is blended and consists of two-thirds reverse osmosis permeate and one-third water treated with activated carbon. The unit cost is $0.28 m^3$, which compares very favorably with the costs of alternative sources of potable water.

West Basin Desalter Project

Historically, the West Ground Water Basin, which is located immediately south and west of the city of Los Angeles, was subjected to heavy overdraft, resulting in seawater intrusion. In the early 1960s, the local water agency, the West Basin Municipal Water District, began to inject good quality water into the basin to act as a seawater intrusion barrier. Although this project succeeded in halting the seawater intrusion, some saline water was trapped inland of the barrier. This saline water continues to degrade ground-water quality, which has TDS of 4,000 mg/l. In 1993 the West Basin district began operating a reverse osmosis facility to reclaim the saline ground waters trapped inland of the seawater intrusion barrier. The facility has a capacity of 5,500 m^3/day and produces potable water supply for the city of Torrance at a cost of $0.42/m^3.

San Luis Rey Basin Desalting Facility

The city of Oceanside, California, which is located midway between Los Angeles and San Diego, depends on a regional water purveyor for domestic water supplies. The need for extensive conveyance facilities to transport this water makes the system vulnerable to a catastrophic earthquake. The system is also vulnerable to extreme drought. The city has no alternative supply of water, and its storage capacity holds only enough supply to meet municipal demands for two days. The San Luis Rey Ground Water Basin, which is adjacent to Oceanside, is salinized, with TDS measuring approximately 1,400 mg/l. In 1994, Oceanside began operation of a 5,570 m^3/day reverse osmosis facility, which desalinates the ground water for potable use. The facility produces water at a reported cost of $0.31/m^3, less than half the cost of a supplemental imported water supply. Product water contains less than 10 mg/l TDS and meets all water quality standards for potable water. The project provides the city with additional water supply and substantial protection against drought or a failure in its conveyance system.

Frances Desalting Facility

Orange County, which lies at the southern margin of the Los Angeles metropolitan area, has historically been short of water and has developed a number of innovative means of augmenting water supplies. One of the local water purveyors, the Irvine Ranch Water District, already provides potable and recycled water to domestic customers through separate plumbing systems, which were installed as part of the planned development of a number of the communities that it serves. The area serviced by the Irvine district continues to grow, and additional sources of water are needed. The underlying ground water is slightly brackish, with a TDS of 870 mg/l. The district has studied various technologies for reclaiming this water, including nanofiltration, electrodialysis, reverse osmosis, and ion exchange. Reverse osmosis was found to be the most cost-effective. The district now proposes to construct a 37,100 m^3/day reverse osmosis ground-water desalting facility. This facility would produce potable water for use within the district's service area and would be the largest ground-water desalination facility in the United States. The cost of the water is estimated to fall in the range of $0.41 to $0.47/m^3. Costs in this range would be substantially less than the costs of alternative sources of potable water.

TABLE 5.13 Costs of Brackish Water Desalination with Reverse
Osmosis

Plant Location	Date On Line	Capacity (m^3/day)	Unit Cost ($/$m^3$)
Water Factory 21, CA	1977	18,670	0.28
Oceanside, CA	1994	7,570	0.43
Arlington Desalter, CA	1990	15,140	0.48
Cape Coral, FL	1976	28,250	0.48
Brighton, CO	1993	10,590	0.50
S.E. Region Water Treatment Facility, IL	1989	5,130	0.62
Sarasota, FL	1982	11,410	0.83
Gasparilla Islands, FL	1990	1,200	0.91
Dare County, NC	1989	5,920	1.05
North Beach, FL	1985	1,140	1.06
Jupiter, FL	1990	7,780	1.09
St. Thomas Dairies, Virgin Islands	1980	57	2.61
Southbay Utilities, FL	1980	472	2.99
Okracoke Sanitary District, NC	1977	395	3.67

NOTE: All costs in 1996 dollars. Total dissolved solids (TDS) of feedwaters are between
1,000 and 5,000 mg/l. TDS of product waters ranges from 25 to 400 mg/l. Operating and
maintenance costs are normalized using an electricity cost of $0.04/kWh. Unit costs com-
puted using a 20 year amortization period for capital costs and a 10 percent interest rate.

SOURCES: Compiled from Cevaal et al., 1995; Kopko et al., 1995; National Water Supply
Improvement Association, 1992; and Szymborski, 1995.

seawater desalination with costs that are relatively attractive compared to
distillation technologies. Table 5.14 shows cost and capacity data for a
selected set of seawater conversion projects employing reverse osmosis
technologies around the world. With the exception of Santa Catalina
Island, CA, where the high cost is particularly attributable to small plant
size and the absence of economies of scale, costs range from approximately $0.90/$m^3$ to $1.35/$m^3$. Thus, although seawater desalting with
reverse osmosis may be less costly than distillation technologies, it is still
quite expensive.

TABLE 5.14 Costs of Seawater Desalination with Reverse Osmosis

Plant Location	Date On Line	Capacity (m³/day)	Unit Cost ($/m³)
Santa Catalina Island, CA[a]	1991	380	3.17
Tigne, Malta[b]	1987	15,140	0.92
Las Palmos, Gran Canaria[b]	1989	35,958	1.35
Rosarita Repowering Project, Mexico[c]	Proposed	37,851	1.24
Jedda I, Phase 1, Saudi Arabia[b]	1989	56,777	0.95

NOTE: All costs are in 1996 dollars. Source water ranges from 36,000 to 47,000 mg/l total dissolved solids. Total dissolved solids for product water are less than 500 mg/l. Operating and maintenance costs are normalized using an electricity cost of $0.04/kWh. Unit costs computed using a 20-year amortization period for capital costs and a 10 percent interest rate.

[a]Modified from National Water Supply Improvement Association, 1992.
[b]Modified from Leitner, 1991.
[c]Modified from Kamal, 1995.

Salinity-Gradient Solar Ponds

A number of researchers have proposed salinity-gradient solar ponds as an alternative means of desalinating seawater. The theoretical advantage of this technology lies in the fact that energy generated from the salinity gradients can be used to power either the desalting process or the desalination facilities.

The pond is maintained with a salinity gradient that increases with depth, with a low-salinity, low-density surface zone floating on top of a high-salinity, high-density lower zone. Between these two layers is a gradient zone of intermediate salinity that acts to isolate the surface and lower zones. The lower zone traps solar energy in the form of heated water. The high density of the lower zone (caused by high salinity) inhibits the heated water from rising, thereby minimizing heat loss to the atmosphere. Lower zone water temperatures can exceed 80°C.

The heat trapped in the lower zone can be used directly to warm saline water for distillation, or indirectly to generate electricity to run desalination facilities. Many researchers have proposed the coupling of salinity-gradient ponds with distillation at temperatures of approximately 70°C and thus the feedwater to a distillation unit could be heated with water from the lower zone of a solar pond. The advantages of the system may include low energy use and low pollution production. Such a technology would be especially suitable for remote arid environments with limited local energy sources. Solar ponds could also be used to generate electricity to operate reverse osmosis desalination plants. However, there are large inefficiencies inherent in producing electricity, compared with directly using water heated in ponds.

Currently, there are no large-scale desalination facilities using salinity-gradient solar ponds. However, a number of pilot studies have been conducted, and some ponds are being used to generate electricity. Since the early 1980s, Israel has operated several salinity-gradient solar ponds for energy production. A pond in Ein Boqek was the first to generate commercial electricity, producing a peak output of 150 kW. Ponds with a total surface area of 250,000 m^2 in Beit Ha'Arava are the heat source for a 5 MW power station. Cost data for proposed solar-pond multieffect distillation facilities are shown in Table 5.15. These data suggest that seawater desalting with solar ponds may sometimes be competitive where the costs of energy are relatively high or energy availability is constrained. The costs do not differ substantially from those of reverse osmosis seawater desalination, whose energy costs are in the range of $0.04 to $0.08/kWh.

The evaluation of seawater desalination is summarized in Table 5.11. Clearly, large-scale seawater desalination could substantially add to the available water supplies of the study area. Seawater conversion using either distillation or membrane technologies is feasible. The environmental impact of seawater conversion mainly relates to the disposition of the concentrated brines resulting from any seawater conversion operation. The method and place of disposal will be important in determining the environmental impact. To date, seawater desalination has only been economically feasible in unique situations where water supplies are intensively constrained. The adverse economics of seawater conversion are

TABLE 5.15 Reported Cost of Seawater Desalination Using Solar Ponds and Distillation Technologies

Reference	Pond Size (1,000 m^2)	Production (m^3/day)	Reported Unit Cost ($/m^3)
Dornon, et al. (1991)	600	9,000	1.13
Glueckstern (1995)	1,200	20,000	0.89
Glueckstern (1995)	12,000	200,000	0.71
Tsilingiris (1995)	35,000	100,000	2.00

likely to persist for the immediate future. The implications for future generations of employing seawater conversion on a large scale are not clear. Certainly, the development of a cost-effective technology that could be passed on would be a positive contribution to future generations. Nevertheless, uncertainties about the economic feasibility of these technologies and their environmental impacts make it difficult to assess their implications for present and future generations.

Cloud Seeding

For many years, people have tried to modify weather to increase water resources. The discovery in the late 1940s that supercooled cloud droplets could be changed to ice crystals by inserting a cooling agent such as dry ice or artificial nuclei such as silver iodide led to cloud seeding to increase precipitation. While cloud seeding has been used in the study area since 1960, its effects on precipitation are still controversial. Moreover, increases in precipitation do not necessarily result in more runoff, which is critical to water supply. The initial optimism about these techniques has been somewhat tempered by the complexities of atmospheric physics (Bruintjes et al., 1992).

From 1961 to 1975, two scientifically designed cloud-seeding experiments were carried out in north and central Israel using a two-target crossover design. The first experiment claimed a positive seeding effect of 15 percent increase in rainfall and the second a positive effect of 13 percent increase in rainfall in the northern part of Israel; the results of both studies were statistically significant at relatively high levels (Nirel and Rosenfeld, 1995). Since 1975, cloud seeding has been used in the north. Increased rainfall from cloud seeding between 1976 and 1990 is estimated at 6 percent, with a 95 percent confidence level (Nirel and Rosenfeld, 1995). The Jordan cloud-seeding program is estimated to have increased rainfall by 19 percent (Tahboub, 1992). Even where the strategy was apparently successful, however, increases in runoff were less significant than increases in precipitation (Benjamini and Harpaz, 1986).

Cloud seeding in the south of Israel has remained experimental. Analyses have not indicated any rainfall increase (Rosenfeld and Farbstein, 1992; Brown et al., 1996). This result may be due to desert dust, possibly from the Sahara-Arabian deserts, which contributes many nuclei and may reduce seeding's effects (Rosenfeld and Farbstein, 1992; Gabriel and Rosenfeld, 1990).

Recent statistical analysis by Rangno and Hobbs (1995) suggests that the cloud-seeding experiments have been compromised by statistical errors, and that neither of the two Israeli experiments demonstrated statistically significant effects on rainfall from cloud seeding. In some circumstances, cloud-seeding methods have been suspected of decreasing precipitation (Rosenfeld et al., 1996). The 1992 World Meteorological Organization Statement on the Status of Weather Modification concluded that "if one were able to precisely predict the precipitation from a cloud system, it would be a simple matter to detect the effect of artificial cloud seeding on that system. The expected effects of seeding are, however, often within the range of natural variability . . . and our ability to predict the natural behavior is still limited" (WMO, 1992).

Since the beginning of the endeavor, there has been international concern about the social and ecological effects of cloud-seeding operations and the economic costs and benefits of the technology (Fleagle et al., 1974). Concerns have also been raised about potential effects on precipitation in downwind countries (WMO, 1992). In 1979, the World Meteorological Organization and the United Nations Environment Program considered draft general guidelines for states concerning weather modification, but they were never finalized (WMO/UNEP, 1979). The guidelines called for notice and consultation with potentially affected countries and for assessments of environmental effects (a point reiterated in the 1992 WMO Statement on Weather Modification). In the 1960s and 1970s, more than a dozen lawsuits were filed in the United States (Brown Weiss, 1983). Further research is still needed to clarify the effects of cloud seeding on precipitation. At this time, it is doubtful that cloud seeding will ever provide a significant source of increased water supply in the study area. Regional cooperation is important to ensure that all countries appreciate the scientific uncertainties about the technology and its impacts. Monitoring for possible effects outside the target area is also important.

Table 5.16 summarizes the evaluation of cloud seeding based on the committee's five criteria. As noted earlier, the effects of cloud seeding on the availability of water in the study area are not completely clear, and it is unlikely that cloud seeding would ever provide a significant source of

TABLE 5.16 Cloud Seeding and Transfers

Committee Criterion	Cloud Seeding	Transfers
1. Impact on Available Water Supply	0–?	0
2. Technically Feasible	+	+/–
3. Environmental Impact	–?	+/–
4. Economically Feasible	?	+/–
5. Implications for Intergenerational Equity	?	?

NOTE: + indicates positive effects, – indicates negative effects, and 0 indicates no impact.

additional water supply. Moreover, very little is known about its technical or economic feasibility or its environmental impact. The implications of cloud seeding for present and future generations are similarly unclear.

Transfers Within the Study Area

Water transfers are used to shift water surpluses generated in one part of the system to another part in need of additional water supplies. There are extensive transfers of water from one area to another within the study region. More transfers are under consideration for the future. In Israel, the National Water Carrier transports an average of 450 million m^3/yr from the north to the south of the country using the Lake Kinneret/ Lake Tiberias/Sea of Galilee as a storage reservoir. In Jordan, the King Abdullah Canal, with a capacity in the northern reaches of 600 million m^3/yr, decreasing to 180 million m^3/yr in the southern part of the valley, and an average flow of 140 to 160 million m^3/yr, diverts water from the Yarmouk River and provides water to irrigate the upper Jordan Valley. The canal also provides the water for the Deir Alla project (consisting of a pipeline, treatment plant, and pumping plants), which provides an average of 35 million m^3/yr to Amman.

The transfer of seawater to the study area has been proposed many times. It would be complex and expensive. The proposed Dead Sea projects would bring saline water from the Mediterranean Sea or Red Sea to the Dead Sea by means of pipelines or canals. The transferred seawater would reverse the trend of falling water levels in the Dead Sea, eventually restoring and maintaining its historic levels; hydroelectric generating stations would take advantage of the elevation difference between the Dead Sea and sea level; and the elevation drop could also furnish the mechanical pressure needed for reverse osmosis desalination, therefore minimizing the use of electricity. The cost-effectiveness of these projects needs to be evaluated based on their total packages of benefits—freshwater production, environmental restoration, continuation of chemical production from the Dead Sea, and power generation.

Transfers of water within the area will not result in any net increase in the available water supply, since they simply reallocate water among uses and places of use. Some transfers will be technically feasible, where facilities exist to move water from one place to another. Other transfers may not be technically feasible, because the physical means to accomplish the transfer are absent. Similarly, the environmental impact of transfers is unclear. Where water is transferred from consumptive uses to environmental uses, the environmental impact is likely to be positive.

Imports of Freshwater into the Study Area

Proposed approaches to import freshwater from outside the region would be complex and expensive, and would require international agreements. They would generally involve moving freshwater by conventional pipelines and canals from other countries such as Turkey (Biswas et al., 1997), or in one instance, transporting the water in large floating plastic bags pulled by tugboats (Tahal Consulting Engineers, 1989). The committee did not evaluate such proposals, because they are outside the scope of this study. Moreover, there is danger that serious consideration of import schemes may prevent the parties in the study area from focusing on the measures that can be taken (such as those described in this study) to provide sustainable water supplies using the region's resources.

CONCLUSIONS

The conventional freshwater sources currently available in the region are barely sufficient to maintain its quality of life and economy. For example, Jordan is currently overexploiting its ground-water resources by about 300 million m^3/yr, thus lowering water levels and creating salinization of freshwater aquifers. Similar examples of overexploitation are occurring throughout the study area. Attempting to meet future regional demands by simply increasing withdrawals of surface and ground water will result in further unsustainable development, characterized by widespread environmental degradation and depletion of freshwater resources. Because these conditions already exist in many parts of the area, for example the Azraq Basin and the Hula Valley, the reality of a constrained water supply is a consideration in formulating government economic plans and policies. Demand and supply can be brought into a sustainable balance only by changing and moderating the pattern of demand by introducing new sources of supply. Above all, water losses should be minimized and water-use efficiency increased substantially. The opportunities offered by specific options to increase and sustain the quantity and quality of the region's freshwater resources are summarized immediately below. Each option deserves careful consideration in terms of practical application and refinement through further research. These options can be initiated in the region within existing legal entitlements to shared water resources.

Conservation

Constraints must be imposed to conserve and limit the use of available water in the study area. By reducing the demand for water, the recommended conservation measures will have a positive effect on water quality and the environment. Voluntary domestic water conservation measures include the following:

- Limiting toilet flushing.
- Adopting water-saving plumbing fixtures, such as toilets and shower heads.
- Adopting water-efficient appliances (notably washing machines).
- Limiting outdoor uses of water, as by watering lawns and gardens during the evening and early morning, and washing cars on lawns and without using a hose.
- Adopting water-saving practices in commerce, such as providing water on request only in restaurants and encouraging multiday use of towels and linens in hotels.
- Repairing household leaks.
- Limiting use of garbage disposal units.

Examples of involuntary domestic water saving measures include the following:

- Repair leaking distribution systems.
- Repair leaking sewer pipes.
- Expand central sewage systems.
- Meter all water connections.
- Ration and restrict water use.

In conclusion, various known methods can lead to significant savings in both indoor and outdoor water use. To implement these methods, government agencies in the study area should consider encouraging their adoption through education, incentives, pricing, taxation, and regulation, and to this end will be involved in setting priorities at various times for the support of needed measures, taking into account the uncertainties attached to the available evidence.

Agriculture

Through rationing, research, and possibly economic pricing policies, agricultural water use can become more efficient. However, as regional nonagricultural water demand increases and the cost of additional water supplies grows more expensive, the role of agriculture in the area's economy will have to be reevaluated, so that as much water as possible is conserved. The region might adopt agricultural practices more in harmony with the ecological realities of drylands. Drylands are and will likely remain marginal for subsistence agriculture, unless the practice is heavily subsidized by water drawn from elsewhere.

A number of useful practices are already used to some degree in the study area, and these practices should be expanded to help conserve agricultural water use:

• Harvesting local water runoff and floodwater to increase water supplies for dryland agriculture.
• Reducing evaporative water loss by cropping within closed environments (desert greenhouses). This method is economic with land and water use, avoids soil salinization, and produces high yields of exportable crops, such as ornamentals, fruits, vegetables, and herbs.
• Using computer-controlled drip "fertigation" (fertilizer applied with irrigation water) and soilless substrates in greenhouses, which economizes on water and fertilizer use and helps prevent ground-water pollution.
• Considering the use of brackish water for irrigation of salinity-tolerant crops.
• Saving more freshwater by switching to irrigation with treated wastewater or with brackish water if possible.
• Changing production from crops with high water requirements to crops with lower water requirements.

Pricing and Pricing Policies

Policies that subsidize the price of water or emphasize revenue recovery to the exclusion of economic efficiency are poorly suited to areas where water is scarce. Conversely, pricing policies that promote economic efficiency and economizing in water use are more appropriate for regions of increasing water scarcity.

Marginal Cost Pricing

The committee recommends the use of marginal cost pricing in the study area to help conserve freshwater resources. As long as marginal costs are higher than average costs, the use of marginal cost pricing will ensure that revenue requirements are met. Marginal cost pricing also sends the correct signals to consumers about the true cost of water and, given some fixed level of benefits, ensures that the costs of providing the water are minimized.

Time-of-Use Pricing

Time-of-use structure discourages use of water during peak-use periods in order to ration water during high use but specifies lower pricing during off-peak usage.

Water Surcharges

Water surcharges, imposed beyond some set level of use, can be employed to discourage excessive use.

Water Markets

Water markets, where marginal cost prices are used, can help allocate water among sectors more efficiently. Markets permit transfers of water to occur on a strictly voluntary basis. Such transfers occur when the difference between the minimum price that sellers are willing to accept and the maximum price that buyers are willing to pay is sufficient to cover any costs of transport or treatment.

Even if water markets are never developed in the study area, simulation of water markets can be very useful in identifying the value of water for alternative uses and regions. Such simulation can also help identify additional water supply and conveyance facilities that are economically justified.

Watershed Management

The concept of total watershed management should be adopted for the study area. This approach has been defined as the art and science of managing the land, vegetation, and water resources of a drainage basin, to control the quality, quantity, and timing of water, toward enhancing and preserving human welfare and nature.

Small Retention Structures and Stormwater Runoff

Small retention structures on the wadis could be effective in capturing stormwater runoff. Stormwater could then be used for artificial recharge of ground water. Urban runoff is another source of water for retention basins. In addition to storing usable water, retention basins would attenuate flooding and avoid excess flows at wastewater treatment plants.

Ground-Water Overdraft

Ground-water mining of an aquifer that is hydraulically connected to a saline water body will deplete the freshwater resource and degrade its quality. An example is the coastal aquifer in Israel and the Gaza Strip, where overexploitation has led to the encroachment of saline water.

Because of the almost immediate environmental consequences of mining aquifers and the later environmental and water quality consequences as well, strong consideration should be given to reducing extraction rates from aquifers in the study area.

To ensure that future generations have sufficient available ground water, research is needed on the amount of water in ground-water storage and the environmental consequences of depleting this storage. In addition, more consideration should be given to the beneficial use of the storage space created by ground-water mining.

Water Harvesting

The region's inhabitants can continue and expand the use of rooftop cisterns for individual domestic supplies. Catchment systems and storage ponds should also be expanded for agricultural water use. Even where conventional sources of water are available, cisterns can provide supplemental water inexpensively and relieve the demand on the water distribution system.

Brackish Water Desalination

Where brackish waters can be desalted, this approach offers a clear promise of augmenting the available water supply. Such desalination is technologically feasible and will not usually have adverse environmental impacts. Economic feasibility depends on the quality of the feedwater, the technology used, and the relative attractiveness of other alternatives.

Underground Dams

On a small scale and under suitable physical conditions, ground-water drainage may be decreased and water levels increased by constructing underground dams. Injection of cement or low-permeability grout through closely spaced boreholes creates a curtain extending to the base of the aquifer. This approach can help prevent lateral salt water intrusion to coastal aquifers.

Wastewater Reclamation

As the demand for water continues to increase beyond the natural supply, it is not unreasonable to forecast a near-total reuse of water in the study area. Reclamation will theoretically double the amount of increased supply brought about by new sources of freshwater as well. Thus, widespread reclamation would decrease the amounts of water needed to meet the probably increased regional demand.

Urban reuse of wastewater requires dual municipal distribution systems—one for potable, the other for reclaimed water. The prospect of major urban expansion in the area provides the incentive to plan communities with an initial dual-water system.

Marginal Quality Water Use

Some savings in freshwater could be obtained by substituting water of marginal quality for some activities now using potable water. But special attention would need to be given to any human health issues when this strategy is under consideration.

Point-of-Use and Point-of-Entry Technologies

Several competing technologies are now available for point-of-use (at the tap) and point-of-entry (at the house) water treatment. These technologies include adsorptive filters, reverse osmosis, ion exchange, and distillation.

Maintenance is essential for these units to function properly. Point-of-entry treatment is a major industry in many countries, supplying potable water to millions of consumers in isolated areas, on farms, and in communities where wells have been contaminated. The method is technically sound and economically feasible for reducing organic and inorganic contaminants. Controlling and monitoring these devices is the key to protecting public health.

REFERENCES

Abu Mayleh, Y. 1991. Hydrological Situation in Gaza Strip. Not published.

Abu Safieh, Y. 1991. Water in the Gaza Strip, The problem and suggestions. In pp. 48-59 of the Proceedings of the Workshop Concerning the Water Situation in the Occupied Territories. Jerusalem, Israel: Hydrology Group.

Al-Kharabsheh, A., R. Al-Weshah, and M. Shatanawi. 1997. Artificial groundwater recharge in the Azraq Basin (Jordan). Dirasat, Agricultural Sciences 24(3)September.

Al-Khodari, R. 1991. The water problem in the Gaza Strip and proposed solutions. In pp. 60-65 of the Proceedings of the Workshop Concerning the Water Situation in the Occupied Territories. Jerusalem, Israel: Hydrology Group.

Al-Weshah, R. A., and D. T. Shaw. 1994. Performance of integrated municipal water systems during drought. ASCE J. Water Resources Planning and Management 120(4)July/August:531-545.

Amiran, D. H. K. 1965. Arid Zone Development: A Reapproval Under Modern Technological Conditions. Economic Geography 41:189-210.

Anonymous. 1988. Collection cisterns. Arab Thought Forum. Sho'on Tanmawiya Maga-zine, pp. 26-28.

Argaman, Y. 1989. Wastewater Reclamation and Reuse in Israel. In Proceedings of the 26th Japan Sewage Works Association, Tokyo.

Assaf, K. 1994. Replenishment of Waters by Artificial Recharge As a Non-Controversial Option in Water Resources Management in the West Bank and Gaza Strip. In Proceedings of the First Israeli-International Academic Conference on Water, Water and Peace in the Middle East, J. Isaac and H. Shuval, eds. Zurich, Switzerland, December, 1992. New York, N.Y.: Elsevier.

Asano, T. 1985. Artificial Recharge of Groundwater. Boston, Mass.: Butterworth Publishers.

Atwater, R., L. Palmquist, and J. Onkka. 1995. The West Basin Desalter: A Viable Alternative. Desalination 103:117-125.

Avnimelech, Y. 1996. Irrigation with wastewater effluents in Israel: The lesson learned. International Conference on Water Resources Management Strategies in the Middle East. November 1996. Society of Water Eng., Israel.

Awerbuch, L. 1988. Desalination Technology: An Overview. Chapter 4. Pp. 53-64 in The Politics of Scarcity: Water in the Middle East, J. R. Starr and D. C. Stoll, eds.. Boulder, Colo.: Westview.

Ayers, R. S., and D. W. Westcot. 1976. Water quality for Agriculture. Irrigation and Drainage Paper 29, pp. 97. Rome: FAO.

Bargouthi, M., and I. Deibes. 1993. Infrastructure and Health Services in the West Bank, Guidelines for Health Care Planning. Health Development Project, West Bank.

Bargur, Y. 1993. Israel Water Economy Toward the First Quarter of the 21st Century: Resource Development Under Deficiency. In Israel 2020. Haifa, Israel: Technion. Pp. 217-265.

Benjamini, Y., and Y. Harpaz. 1986. Observational Rainfall-runoff Analysis for Estimating Effects of Cloud Seeding on Water Resources in Northern Israel. J. Hydrology 83(3/4):299-306.

Binyamin, Y., S. Marish, A Gafni, and M. Gutman. 1991. Subsurface Drainage Systems as a Means for Decreasing Salinity, Final Report (in Hebrew). Division of Soil Conservation, Israel Ministry of Agriculture.

Biswas, A. K., J. Kolars, M. Murakami, J. Waterbury, and A. Wolf. 1997. Core and Periphery: A Comprehensive Approach to Middle Eastern Water. Middle East Water Commission. Delhi.: Oxford University Press.

Boland, J. J. 1993. Pricing Urban Water: Principles and Compromises. Water Resources Update (92):7-10.

BRL-ANTEA. 1995. Guidelines for a Master Plan for Water Management in the Jordan River Basin. Ingenierie, France: BRL. Not published.

Brown, D., Y. Bechar, and K. Pick. 1996. Rain Enhancement Operations in Israel. 13th Conference on Weather Modification. American Meteorological Society.

Brown Weiss, E. 1983. Management of Weather and Climate Disputes. UCLA Journal of Environmental Law and Policy 3:275.

Bruintjes, R. T., T. L. Clark, and W. D. Hall. 1992. The Present Status of Precipitation Enhancement by Cloud Seeding. Pp. 612-617 in Proceedings of Irrigation and Drainage Sessions, Water Forum '92. Baltimore, Md: ASCE.

Burke, M. E., and G. A. Stasko. 1987. Organizing water quality districts in New York State. J. American Water Works Association October:39-41.

Calderon, R. L., A. F. Dufour, and E. W. Mood. 1987. Significance of bacteria associated with point-of-use granulated activated carbon filters. Presented at Water Quality Association Conference, Dallas, Texas. March.

Carter, H. O., H. J. Vaux, Jr., and A. F. Scheuring, eds. 1994. Sharing Scarcity: Gainers and Losers in Water Marketing. University of California, Agricultural Issues Center, Davis, California.

CES Consulting Engineers and GTZ. 1996. Middle East Regional Study on Water Supply and Demand Development, Phase I, Regional Overview. Sponsored by the Government of the Federal Republic of Germany for the Multilateral Working Group on Water Resources. Association for Technical Cooperation (GTZ), Eschborn, Germany.

Cevaal, J. N., W. B. Suratt, and J. E. Burke. 1995. Nitrate Removal and Water Quality: Improvements with Reverse Osmosis for Brighton, Colorado. Desalination 103:101-111.

Crook, J. 1990. Water Reclamation. Pp. 157-187 in Encyclopedia of Physical Science and Technology, R. Myers, ed. San Diego, Calif.: Academic Press, Inc.

Dorman, B., J. Weinberg, and U. Fisher. 1991. Solar Pond as an Actual Solution for Desalination. International Desalination Association World Conference on Desalination and Reuse. Washington, D.C.

Dunivin, W., P. H. Lange, R. G. Sudak, and M. Wilf. 1991. Reclamation of Ground Water Using RO Technology. International Desalination Association World Conference on Desalination and Water Reuse, Washington, D.C.

East Bay Municipal Utilities District. 1992. 1992 Annual Report. Office of Reclamation. Oakland, CA.

Eitan, G. 1995. Survey of Collection, Treatment, and Utilization of Effluent. Ministry of Agriculture, Israel Water Commission. 175 pp.

Evenari, M., L. Shanan, and N. Tadmor. 1982. The NEGEV. The Challenge of a Desert. Cambridge, Mass.: Harvard Univ. Press.

Everest, W. R., and J. T. Morris. 1995. The Frances Desalter: A Key to Water Independence on the Irvine Ranch. Desalination 103:127-132.

Fatafta, A., F. El-Karmi, N. Al-Haj Ali, K. Al-Rawajfeh, F. Abu Niaj, F. Anani, M. Bseiso, and N. Halaseh. 1992. Water Desalination Technology in Jordan. Amman, Jordan: Higher Council for Science and Technology. 110 p.

Filteau, G., C. Whitley, and I. C. Watson. 1995. Water Reclamation Fuels Economic Growth in Harlingen, Texas: Reclamation of Municipal Wastewater for Industrial Processing Use. Desalination 103:31-37.

Fisher, F. M., N. Harshadeep, A. Nevo, et. al. 1996. The Economics of Water: An Application to the Middle East. Unpublished Manuscript. March 25, 1996.

Fleagle, R. G., J. A. Crutchfield, R. W. Johnson, and M. F. Abdo. 1974. Weather Modification in the Public Interest. American Meteorological Society and University of Washington Press. 88 pp.

Gabriel, K. R., and D. Rosenfeld. 1990. The Second Israeli Rainfall Stimulation Experiment: Analysis of Precipitation on Both Targets. J. Applied Meteorology 29(10):1055-1067.

Garagunis, C. N. 1981. Construction of an Impervious Diaphragm for Improvement of a Subsurface Water-Reservoir and Simultaneous Protection from Migrating Salt Water. Bull. International Association of Engineering Geology 24:169-172.

Glueckstern, P. 1991. Cost Estimates of Large RO Systems. Desalination 81:49-56.

Glueckstern, P. 1995. Potential Uses of Solar Energy for Seawater Desalination. Desalination 101:47-50.

Goldberg, S. 1992. Israel's natural water resources potential-quality, quantity, dependability. In the water economy of Israel workshop No. 148. Rehwat, Israel: The Center for Agricultural Economics Research (Hebrew).

Grattan, S. R., and J. D. Rhoads. 1990. Irrigation with Saline Ground Water and Drainage Water. In Agricultural Salinity Assessment and Management, K. K. Tanji, ed.

Grebbien, V. L. 1991. West Basin Municipal Water District. International Desalination Association World Conference on Desalination and Reuse. Washington, D.C.

Gupta, A., and M. Chaudri. 1992. Domestic water purification for developing countries. AQUA J. Water Supply Research and Technology 41(5)October.

Heitmann. H.-G., ed. 1990. Saline Water Processing: Desalinization and Functions of Sewater, Brackish water, and Irrigation Wastewater. Weinheim, Federal Republic of Germany. New York.

Hirshleifer, J., J. C. DeHaven, and J. Milliman. 1960. Water Supply: Economics, Technology and Policy. Chicago, Ill.: University of Chicago Press.

Hoffman, D. 1994. Potential Applications for Desalination in the Area. Pp. 315-327 in Proceedings of the First Israeli-International Academic Conference on Water, Water and Peace in the Middle East, J. Isaac and H. Shuval, eds., Zurich, Switzerland, December, 1992. New York, N.Y.: Elsevier.

Ishaq, A. M., and H. A. Khararjian. 1988. Stormwater Harvesting in the Urban Watersheds of Arid Zones. Water Resources Bulletin 24(6):1227-1235.

Juanico, M. 1993. Alternative schemes for municipal sewage treatment and disposal in industrial countries: Israel as a case study. Ecological Eng. 2:101-118.

Juanico, M. 1996. The performance of batch stabilization reservoirs for wastewater treatment, storage and reuse in Israel. Water Sci. Technol. 33:149-159.

Kamal, I. 1995. An Assessment of Desalination Technology for the Rosarito Repowering Project. Desalination 102:269-278.

Kanarek, A., A. Aharoni, M. Michail, I. Kogan, and D. Sherer. 1994. Dan Region Reclamation Project, Groundwater Recharge with Municipal Effluent. Tel Aviv, Israel: Mekorot Water Company Ltd. 150 pp. (Hebrew).

Kifaya, B. 1991. In Salameh and Garber: Water Resources of Jordan and their Future Potentials. Friedrich Ebert Foundation, Higher Council of Science and Technology, Amman, Jordan.

Krishna, J. H. 1991. Improving Cistern Water Quality. Proceedings of the Fifth International Conference on Rain Water Cistern Systems, Keelung, Taiwan.

Kopko, S., M. Seamans, J. E. Nemeth, and I. C. Watson. 1995. Desalting in Cape Coral, FL: An Operating Update. Desalination 102:245-253.

Leitner, G. F. 1991. Seawater Desalting by RO: What Does It Cost? International Desalination Association World Conference on Desalination and Reuse, Washington, D.C.

Ministry of Agriculture, Water Commission, Hydrology Service. 1967. Hydrology Year Book of Israel, Jerusalem.

Natarajan, R., W. V. B. Ramalingam, and W. P. Harkare. 1991. Experience in Installation and Operation of Brackish Water Desalination Plants in Rural Areas of India. International Desalination Association World Conference on Desalination and Water Reuse, Washington, D.C.

National Research Council (NRC). 1992. Water Transfers in the West: Efficiency, Equity and the Environment. Washington, D.C.: National Academy Press.

National Research Council (NRC). 1994. Ground Water Recharge Using Waters of Impaired Quality. Washington, D.C.: National Academy Press. 283 pp.

National Research Council (NRC). 1995. Mexico City's Water Supply—Improving the Outlook for Sustainability. Washington, D.C.: National Academy Press. 107 pp.

National Water Supply Improvement Association. 1992. Potable Water Desalination in the U.S., Capital Costs, Operating Costs and Water Selling Prices. Sacramento, Calif.: American Desalting Association. 61 pp.

Nilsson, A. 1988. Groundwater Dams for Small-scale Water Supply. London, U.K.: Intermediate Technology Publications Ltd. 69 pp.

Nirel, R., and D. Rosenfeld. 1995. Estimation of the Effect of Operational Seeding on Rain Amounts in Israel. J. of Applied Meteorology 34:2220-2229.

Okun, D. A. 1994. The Role of Reclamation and Reuse in Addressing Community Water Needs in Israel and the West Bank. Pp. 329-338 in Proceedings of the First Israeli-International Academic Conference on Water, Water and Peace in the Middle East, J. Isaac and H. Shuval, eds. Zurich, Switzerland, December, 1992. New York, N.Y.: Elsevier.

Palestinian Water Authority - Ministry of Planning and International Cooperation (MOPIC). 1996. Gaza Water Resources, Policy Direction in Ground Water Protection and Pollution Control. Gaza.

Palestinian Hydrology Group (PHG). 1992. Rural Areas Water Needs Assessment. Not published. Jerusalem.

Palestinian Hydrology Group (PHG). 1995. Three Year Plan, Water Resources Rehabilitation and Development in the West Bank and Gaza. Not published. Jerusalem.

Planning and Research Center. 1993. Population Handbook. Quoted in Al-Fajr Newspaper, 10 May 1993, Jerusalem.

Rangno, A. L., and P. V. Hobbs. 1995. A New Look at the Israeli Cloud Seeding Experiments. J. Applied Meteorology 34:1169-1193.

Rosenfeld, D., and H. Farbstein. 1992. Possible Influence of Desert Dust on Seedability of Clouds in Israel. J. of Applied Meteorology 31:722-31.

Rosenfeld, D., I. L. Sednev, and A. P. Khain. 1996. Numerical Simulation of Cloud Effects Using a Spectral Microphysics Cloud Model. 13th Conference on Weather Modification. American Meteorological Society.

Rozelle, L. T. 1987. Point-of-use and point-of-entry drinking water treatment. J. American Water Works Association 79(10)October.

Russell, C. S., and B. S. Shin. 1996. Public Utility Priding: Theory and Practical Limitations. Pp. 123-139 in Marginal Cost Rate Design and Wholesale Water Markets, D. Hall, ed. Greenwich, Conn.: JAI Press, Inc.

Rymon, D., and U. Or. 1991. An example of improved water use efficiency in traditional farming: The case of the Jiftlik Valley. Sustainable Agric. 2:103-118.

Salameh, E., and H. Bannayan. 1993. Water resources of Jordan—Present Status and Future Potentials. Amman, Jordan: Friedrich Ebert Stiftung. 183 pp.

Sbeih, M. Y. 1994. Reuse of Waste Water for Irrigation in the West Bank: Some Aspects. Pp. 339-350 in Proceedings of the First Israeli- International Academic Conference on Water, Water and Peace in the Middle East, J. Isaac and H. Shuval, eds., Zurich, Switzerland, December, 1992. New York, N.Y.: Elsevier.

Sexton, R., et al. 1989. The Conservation and Welfare effects of information in a time-of-day pricing experiment. Land Economics 65 (3):272-279.

Shalhevet, J. 1994. Using Water of Marginal Quality for Crop Production. Agricultural Water Management 25:233-269.

Shalhevet, J., A. Mantell, H. Bielori, and D. Shimshi. 1981. Irrigation of Field and Orchard Crops under Semi-arid Conditions. IIIC Publication No. 1, 132 pp.

Shelef, G. 1991. Wastewater reclamation and water resources management. Wastewater Reclamation and Reuse 24:251-265.

Soffer, A. 1992. Rivers of Fire—Conflict of Water in the Middle East. Tel Aviv, Israel: Am oved Publ. 258 pp.

Stanhill, G. 1992. Irrigation in Israel: Past achievements, present challenges and future possibilities. Pp. 63-77 in Water Use Efficiency in Agriculture, J. Shalhevet, C. M. Liu, and Y. X. Xu, eds. Rehovot, Israel: Priel Publishers.

Szymborski, S. E. 1995. Early Action Plan—San Luis Rey Desalting Facility, City of Oceanside, California. Desalination 103:147-153.

Tahal Consulting Engineers. 1989. Supply of Water by Sea from Turkey to Israeli Prefeasibility Study—Draft Version. Tel Aviv, Israel: Tahal Consulting Engineers.

Tahal Consulting Engineers. 1993. Israel water study for the World Bank. Working Paper. Tel Aviv, Israel: Tahal Consulting Engineers. March.

Tahboub, I. K. 1992. Status of the Precipitation Enhancement Program in Jordan. Amman, Jordan: Meteorological Department, Water Authority of Jordan. 54 pp.

Tekeli, S., and M. H. Mahmood. 1987. Application of Roof Rainwater Cistern Systems in Jordan. Proceedings of the Third International Conference on Rain Water Cistern Systems, Khon Kaen University, Thailand.

Tobin, R. S. 1987. Testing and evaluating point-of-use treatment devices in Canada. J. American Water Works Association, October.

Tsilingiris, P. T. 1995. The Analysis and Performance of Large-Scale Stand-Alone Solar Desalination Plants. Desalination 103:240-255.

U.S. Army Corps of Engineers. 1979. Report on Water Resources Study, Phase II, and Technical Proposal for Construction of Water Recharge Projects. Oman Ministry of Agriculture, Fisheries, Petroleum and Minerals. 25 pp.

U.S. Environmental Protection Agency (EPA). 1992. Guidelines for Water Reuse. EPA Manual. Washington, D.C.: U. S. Environmental Protection Agency. 247 pp.

U.S. Environmental Protection Agency (EPA). 1995. Cleaner Water Through Conservation. Office of Water, EPA 841-B-95-002. Washington, D.C.: U.S. Environmental Protection Agency.

Wachtel, B. 1994. The Peace Canal Project: A Multiple Conflict Resolution Perspective for the Middle East. Pp. 363-374 in Proceedings of the First Israeli-Palestinian International Academic Conference on Water, Water and Peace in the Middle East, J. Isaac and H. Shuval, eds., Zurich, Switzerland, December, 1992. New York, N.Y.: Elsevier.

Wagnick Consutling GmbH. 1996. 1996 IDA Worldwide Desalting Inventory, Report No. 14. Gnarrenburg, Germany.

World Meteorological Organization (WMO). 1992. WMO Statement on the Status of Weather Modification. Approved July 1992.

WMO/UNEP (World Meteorological Organization, United Nations Environment Programme). 1979. Report of WMO/UNEP Meeting of Experts Designated by Governments on the Legal Aspects of Weather Modification.

Yaziz, M. I., H. Gunting, N. Sapari, and A.W. Ghazali. 1989. Variations in Rainwater Quality from Roof Catchments. Water Research 23(6):761-765.

Appendixes

A
Excerpts from the Treaty of Peace Between the State of Israel and the Hashemite Kingdom of Jordan October 26, 1994

ARTICLE 6
WATER

With the view to achieving a comprehensive and lasting settlement of all the water problems between them:

1. The Parties agree mutually to recognise the rightful allocations of both of them in Jordan River and Yarmouk River waters and Araba/Arava ground water in accordance with the agreed acceptable principles, quantities and quality as set out in Annex II, which shall be fully respected and complied with.

2. The Parties, recognising the necessity to find a practical, just and agreed solution to their water problems and with the view that the subject of water can form the basis for the advancement of cooperation between them, jointly undertake to ensure that the management and development of their water resources do not, in any way, harm the water resources of the other Party.

3. The Parties recognise that their water resources are not sufficient to meet their needs. More water should be supplied for their use through various methods, including projects of regional and international co-operation.

4. In light of paragraph 3 of this Article, with the understanding that co-operation in water-related subjects would be to the benefit of both Parties, and will help alleviate their water shortages, and that water issues along their entire boundary must be dealt with in their totality, including the possibility of trans-boundary water transfers, the Parties agree to search for ways to alleviate water shortage and to co-operate in the following fields:

   a. development of existing and new water resources, increasing the water availability including cooperation on a regional basis as appropriate, and minimising wastage of water resources through the chain of their uses;

   b. prevention of contamination of water resources;

   c. mutual assistance in the alleviation of water shortages;

   d. transfer of information and joint research and development in water-related subjects, and review of the potentials for enhancement of water resources development and use.

5. The implementation of both Parties' undertakings under this Article is detailed in Annex II.

ISRAEL-JORDAN PEACE TREATY
ANNEX IV
ENVIRONMENT

Israel and Jordan acknowledge the importance of the ecology of the region, its high environmental sensitivity and the need to protect the environment and prevent danger and risks for the health and well-being of the region's population. They both recognise the need for conservation of natural resources, protection of biodiversity and the imperative of attaining economic growth based on sustainable development principles.

In light of the above, both Parties agree to co-operate in matters relating to environmental protection in general and to those that may mutually effect them. Areas of such co-operation are detailed as follows:

A. Taking the necessary steps both jointly and individually to prevent damage and risks to the environment in general, and in particular those that may affect people, natural resources and environmental assets in the two countries respectively.

B. Taking the necessary steps by both countries to cooperate in the following areas:

- Environmental planning and management including conducting Environmental Impact Assessment (EIA) and exchanging of data on projects possessing potential impact on their respective environments.

- Environmental legislation, regulations, standards and enforcement thereof.

- Research and applied technology.

- Emergency response, monitoring, related notification procedures and control of damages.

- Code of conduct through regional charters.

- This may be achieved through the establishment of joint modalities and mechanisms of cooperation to ensure exchange of information, communication and coordination regarding matters and activities of mutual environmental concern between their environmental administrations and experts.

C. Environmental subjects to be addressed:

1. Protection of nature, natural resources and biodiversity, including cooperation in planning and management of adjacent protected areas along the common border, and protection of endangered species and migratory birds.

2. Air quality control, including general standards, criteria and all types of man-made hazardous radiations, fumes and gases.

3. Marine environment and coastal resources management.

4. Waste management including hazardous wastes.

5. Pest control including house flies and mosquitoes, and prevention of diseases transferred by pests, such as malaria and leishmaniosis.

6. Abatement and control of pollution, contamination and other manmade hazards to the environment.

7. Desertification: combatting desertification, exchange of information and research knowledge, and the implementation of suitable technologies.

8. Public awareness and environmental education, encouraging the exchange of knowledge, information, study materials, education programmes and training through public actions and awareness campaigns.

9. Noise: reducing noise pollution through regulation, licensing and enforcement based on agreed standards.

10. Potential cooperation in case of natural disasters.

D. In accordance with the above, the two Parties agree to cooperate in activities and projects in the following geographical areas:

I. The Gulf of Aqaba

I.1 The Marine Environment:

- Natural resources.
- Coastal reef protection.
- Marine pollution:

  —Marine sources: such as oil spills, littering and waste disposal and others.
  —Land-based sources: such as liquid waste, solid waste and littering.
  —Abatement, including monitoring and emergency response actions.

I.2 Coastal Zone Management - The Littoral

- Nature reserves and protected areas.
- Environmental protection of water resources.
- Liquid waste.
- Solid waste.
- Tourism and recreational activities.
- Ports.

- Transport.
- Industry and power generation.
- Air quality.
- Hazardous materials.
- Environmental assessments.

II.The Rift Valley

II.1 The Jordan River

Israel and Jordan agree to cooperate along the common boundaries in the following aspects:

- Ecological rehabilitation of the Jordan River.
- Environmental protection of water resources to ensure optimal water quality, at reasonably usable standards.
- Agricultural pollution control.
- Liquid waste.
- Pest control.
- Nature reserves and protected areas.
- Tourism and historical heritage.

II.2 The Dead Sea

- Nature reserves and protected areas.
- Pest control.
- Environmental protection of water resources.
- Industrial pollution control.
- Tourism and historical heritage.

II. Emek Ha'arava/Wadi Araba

- Environmental protection of water resources.
- Nature reserves and protected areas.
- Pest control.
- Tourism and historical heritage.
- Agricultural pollution control.

ISRAEL-JORDAN PEACE TREATY
ANNEX II
WATER RELATED MATTERS

Pursuant to Article 6 of the Treaty, Israel and Jordan agreed on the following Articles on water related matters.

Article I: Allocation

1. Water from the Yarmouk River

   a. Summer period - 15th May to 15th October of each year. Israel pumps (12) MCM and Jordan gets the rest of the flow.

   b. Winter period - 16th October to 14th May of each year. Israel pumps (13) MCM and Jordan is entitled to the rest of the flow subject to provisions outlined hereinbelow: Jordan concedes to Israel pumping an additional (20) MCM from the Yarmouk in winter in return for Israel conceding to transferring to Jordan during the summer period the quantity specified in paragraphs (2.a) below from the Jordan River.

   c. In order that waste of water will be minimized, Israel and Jordan may use, downstream of point 121/Adassiya Diversion, excess flood water that is not usable and will evidently go to waste unused.

2. Water from the Jordan River

   a. Summer period - 15th May to 15th October of each year. In return for the additional water that Jordan concedes to Israel in winter in accordance with paragraph (1.b) above, Israel concedes to transfer to Jordan in the summer period (20) MCM from the Jordan River directly upstream from Deganya gates on the river. Jordan shall pay the operation and maintenance cost of such transfer through existing systems (not including capital cost) and shall bear the total cost of any new transmission system. A separate protocol shall regulate this transfer.

   b. Winter period - 16th October to 14th May of each year. Jordan is entitled to store for its use a minimum average of (20) MCM of the floods in the Jordan River south of its confluence with the Yarmouk (as outlined in Article II below). Excess floods that are not usable and that will otherwise be wasted can be utilised for the benefit of the two Parties including pumped storage off the course of the river.

c. In addition to the above, Israel is entitled to maintain its current uses of the Jordan River waters between its confluence with the Yarmouk and its confluence with Tirat Zvi/Wadi Yabis. Jordan is entitled to an annual quantity equivalent to that of Israel, provided however, that Jordan's use will not harm the quantity or quality of the above Israeli uses. The Joint Water Committee (outlined in Article VII below) will survey existing uses for documentation and prevention of appreciable harm.

d. Jordan is entitled to an annual quantity of (10) MCM of desalinated water from the desalination of about (20) MCM of saline springs now diverted to the Jordan River. Israel will explore the possibility of financing the operation and maintenance cost of the supply to Jordan of this desalinated water (not including capital cost). Until the desalination facilities are operational, and upon the entry into force of the Treaty, Israel will supply Jordan (10) MCM of Jordan River water from the same location as in (2.a) above, outside the summer period and during dates Jordan selects, subject to the maximum capacity of transmission.

3. Additional Water

Israel and Jordan shall cooperate in finding sources for the supply to Jordan of an additional quantity of (50) MCM/year of water of drinkable standards. To this end, the Joint Water Committee will develop, within one year from the entry into force of the Treaty, a plan for the supply to Jordan of the abovementioned additional water. This plan will be forwarded to the respective governments for discussion and decision.

4. Operation and Maintenance

a. Operation and maintenance of the systems on Israeli territory that supply Jordan with water, and their electricity supply, shall be Israel's responsibility. The operation and maintenance of the new systems that serve only Jordan will be contracted at Jordan's expense to authorities or companies selected by Jordan.

b. Israel will guarantee easy unhindered access of personnel and equipment to such new systems for operation and maintenance. This subject will be further detailed in the agreements to be signed between Israel and the authorities or companies selected by Jordan.

Article II: Storage

1. Israel and Jordan shall cooperate to build a diversion/storage dam on the Yarmouk River directly downstream of the point 121/Adassiya Diversion. The purpose is to improve the diversion efficiency into the King Abdullah Canal of the water allocation of the Hashemite Kingdom of Jordan, and possibly for the diversion of Israel's allocation of the river water. Other purposes can be mutually agreed.

2. Israel and Jordan shall cooperate to build a system of water storage on the Jordan River, along their common boundary, between its confluence with the Yarmouk River and its confluence with Tirat Zvi/ Wadi Yabis, in order to implement the provision of paragraph (2.b) of Article I above. The storage system can also be made to accommodate more floods; Israel may use up to (3) MCM/year of added storage capacity.

3. Other storage reservoirs can be discussed and agreed upon mutually.

Article III: Water Quality and Protection

1. Israel and Jordan each undertake to protect, within their own jurisdiction, the shared waters of the Jordan and Yarmouk Rivers, and Arava/ Araba groundwater, against any pollution, contamination, harm or unauthorized withdrawals of each other's allocations.

2. For this purpose, Israel and Jordan will jointly monitor the quality of water along their boundary, by use of jointly established monitoring stations to be operated under the guidance of the Joint Water Committee.

3. Israel and Jordan will each prohibit the disposal of municipal and industrial wastewater into the course of the Yarmouk or the Jordan Rivers before they are treated to standards allowing their unrestricted

agricultural use. Implementation of this prohibition shall be completed within three years from the entry into force of the Treaty.

4. The quality of water supplied from one country to the other at any given location shall be equivalent to the quality of the water used from the same location by the supplying country.

5. Saline springs currently diverted to the Jordan River are earmarked for desalination within four years. Both countries shall cooperate to ensure that the resulting brine will not be disposed of in the Jordan River or in any of its tributaries.

6. Israel and Jordan will each protect water systems in its own territory, supplying water to the other, against any pollution, contamination, harm or unauthorised withdrawal of each other's allocations.

Article IV: Groundwater in Emek Ha'arava/Wadi Araba

1. In accordance with the provisions of this Treaty, some wells drilled and used by Israel along with their associated systems fall on the Jordanian side of the borders. These wells and systems are under Jordan's sovereignty. Israel shall retain the use of these wells and systems in the quantity and quality detailed an Appendix to this Annex, that shall be jointly prepared by 31st December, 1994. Neither country shall take, nor cause to be taken, any measure that may appreciably reduce the yields of quality of these wells and systems.

2. Throughout the period of Israel's use of these wells and systems, replacement of any well that may fail among them shall be licensed by Jordan in accordance with the laws and regulations then in effect. For this purpose, the failed well shall be treated as though it was drilled under license from the competent Jordanian authority at the time of its drilling. Israel shall supply Jordan with the log of each of the wells and the technical information about it to be kept on record. The replacement well shall be connected to the Israeli electricity and water systems.

3. Israel may increase the abstraction rate from wells and systems in Jordan by up to (10) MCM/year above the yields referred to in paragraph 1 above, subject to a determination by the Joint Water Committee that this undertaking is hydrogeologically feasible and does not

harm existing Jordanian uses. Such increase is to be carried out within five years from the entry into force of the Treaty.

4. Operation and Maintenance

    a. Operation and maintenance of the wells and systems on Jordanian territory that supply Israel with water, and their electricity supply shall be Jordan's responsibility. The operation and maintenance of these wells and systems will be contracted at Israel's expense to authorities or companies selected by Israel.

    b. Jordan will guarantee easy unhindered access of personnel and equipment to such wells and systems for operation and maintenance. This subject will be further detailed in the agreements to be signed between Jordan and the authorities or companies selected by Israel.

Article V: Notification and Agreement

1. Artificial changes in or of the course of the Jordan and Yarmouk Rivers can only be made by mutual agreement.

2. Each country undertakes to notify the other, six months ahead of time, of any intended projects which are likely to change the flow of either of the above rivers along their common boundary, or the quality of such flow. The subject will be discussed in the Joint Water Committee with the aim of preventing harm and mitigating adverse impacts such projects may cause.

Article VI: Cooperation

1. Israel and Jordan undertake to exchange relevant data on water resources through the Joint Water Committee.

2. Israel and Jordan shall co-operate in developing plans for purposes of increasing water supplies and improving water use efficiency, within the context of bilateral, regional or international cooperation.

Article VII: Joint Water Committee

1. For the purpose of the implementation of this Annex, the Parties will
    establish a Joint Water Committee comprised of three members from
    each country.

2. The Joint Water Committee will, with the approval of the respective
    governments, specify its work procedures, the frequency of its meet-
    ings, and the details of its scope of work. The Committee may invite
    experts and/or advisors as may be required.

3. The Committee may form, as it deems necessary, a number of special-
    ized sub-committees and assign them technical tasks. In this context,
    it is agreed that these sub-committees will include a northern sub-
    committee and a southern sub-committee, for the management on the
    ground of the mutual water resources in these sectors.

B

Excerpts from the Israeli-Palestinian Interim Agreement on the West Bank and Gaza Strip September 28, 1995

ARTICLE 40
Water and Sewage

On the basis of good-will, both sides have reached the following agreement in the sphere of Water and Sewage.

Principles

1. Israel recognizes the Palestinian water rights in the West Bank. These will be negotiated in the permanent status negotiations and settled in the Permanent Status Agreement relating to the various water resources.

2. Both sides recognize the necessity to develop additional water for various uses.

3. While respecting each side's powers and responsibilities in the sphere of water and sewage in their respective areas, both sides agree to coordinate the management of water and sewage resources and systems in the West Bank during the interim period, in accordance with the following principles:

   a. Maintaining existing quantities of utilization from the resources, taking into consideration the quantities of additional water for the Palestinians from the Eastern Aquifer and other agreed sources in the West Bank as detailed in this Article.

b. Preventing the deterioration of water quality in water resources.

c. Using the water resources in a manner which will ensure sustainable use in the future, in quantity and quality.

d. Adjusting the utilization of the resources according to variable climatological and hydrological conditions.

e. Taking all necessary measures to prevent any harm to water resources, including those utilized by the other side.

f. Treating, reusing or properly disposing of all domestic, urban, industrial, and agricultural sewage.

g. Existing water and sewage systems shall be operated, maintained and developed in a coordinated manner, as set out in this Article.

h. Each side shall take all necessary measures to prevent any harm to the water and sewage systems in their respective areas.

i. Each side shall ensure that the provisions of this Article are applied to all resources and systems, including those privately owned or operated, in their respective areas.

Transfer of Authority

4. The Israeli side shall transfer to the Palestinian side, and the Palestinian side shall assume, powers and responsibilities in the sphere of water and sewage in the West Bank related solely to Palestinians, that are currently held by the military government and its Civil Administration, except for the issues that will be negotiated in the permanent status negotiations, in accordance with the provisions of this Article.

5. The issue of ownership of water and sewage related infrastructure in the West Bank will be addressed in the permanent status negotiations.

Additional Water

6.   Both sides have agreed that the future needs of the Palestinians in the West Bank are estimated to be between 70-80 mcm/year.

7.   In this framework, and in order to meet the immediate needs of the Palestinians in fresh water for domestic use, both sides recognize the necessity to make available to the Palestinians during the interim period a total quantity of 28.6 mcm/year, as detailed below:

   a.   Israeli Commitment:

      (1)   Additional supply to Hebron and the Bethlehem area, including the construction of the required pipeline—1 mcm/year.

      (2)   Additional supply to Ramallah area—0.5 mcm/year.

      (3)   Additional supply to an agreed take-off point in the Salfit area—0.6 mcm/year.

      (4)   Additional supply to the Nablus area—1 mcm/year.

      (5)   The drilling of an additional well in the Jenin area—1.4 mcm/year.

      (6)   Additional supply to the Gaza Strip—5 mcm/year.

      (7)   The capital cost of items (1) and (5) above shall be borne by Israel.

   b.   Palestinian Responsibility:

      (1)   An additional well in the Nablus area—2.1 mcm/year.

      (2)   Additional supply to the Hebron, Bethlehem and Ramallah areas from the Eastern Aquifer or other agreed sources in the West Bank—17 mcm/year.

      (3)   A new pipeline to convey the 5 mcm/year from the existing Israeli water system to the Gaza Strip. In the future, this quantity will come from desalination in Israel.

      (4)   The connecting pipeline from the Salfit take-off point to Salfit.

    (5) The connection of the additional well in the Jenin area to the consumers.

    (6) The remainder of the estimated quantity of the Palestinian needs mentioned in paragraph 6 above, over the quantities mentioned in this paragraph (41.4-51.4 mcm/year), shall be developed by the Palestinians from the Eastern Aquifer and other agreed sources in the West Bank. The Palestinians will have the right to utilize this amount for their needs (domestic and agricultural).

8. The provisions of paragraphs 6-7 above shall not prejudice the provisions of paragraph 1 to this Article.

9. Israel shall assist the Council in the implementation of the provisions of paragraph 7 above, including the following:

    a. Making available all relevant data.

    b. Determining the appropriate occasions for drilling of wells.

10. In order to enable the implementation of paragraph 7 above, both sides shall negotiate and finalize as soon as possible a Protocol concerning the above projects, in accordance with paragraphs 18 - 19 below.

The Joint Water Committee

11. In order to implement their undertakings under this Article, the two sides will establish, upon the signing of this Agreement, a permanent Joint Water Committee (JWC) for the interim period, under the auspices of the CAC.

12. The function of the JWC shall be to deal with all water and sewage related issues in the West Bank including, *inter alia*:

    a. Coordinated management of water resources.

    b. Coordinated management of water and sewage systems.

    c. Protection of water resources and water and sewage systems.

d. Exchange of information relating to water and sewage laws and regulations.

e. Overseeing the operation of the joint supervision and enforcement mechanism.

f. Resolution of water and sewage related disputes.

g. Cooperation in the field of water and sewage, as detailed in this Article.

h. Arrangements for water supply from one side to the other.

i. Monitoring systems. The existing regulations concerning measurement and monitoring shall remain in force until the JWC decides otherwise.

j. Other issues of mutual interest in the sphere of water and sewage.

13. The JWC shall be comprised of an equal number of representatives from each side.

14. All decisions of the JWC shall be reached by consensus, including the agenda, its procedures and other matters.

15. Detailed responsibilities and obligations of the JWC for the implementation of its functions are set out in Schedule 8.

Supervision and Enforcement Mechanism

16. Both sides recognize the necessity to establish a joint mechanism for supervision over and enforcement of their agreements in the field of water and sewage, in the West Bank.

17. For this purpose, both sides shall establish, upon the signing of this Agreement, Joint Supervision and Enforcement Teams (JSET), whose structure, role, and mode of operation is detailed in Schedule 9.

Water Purchases

18. Both sides have agreed that in the case of purchase of water by one side from the other, the purchase shall pay the full real cost incurred by the supplier, including the cost of production at the source and the

conveyance all the way to the point of delivery. Relevant provisions will be included in the Protocol referred to in paragraph 19 below.

19. The JWC will develop a Protocol relating to all aspects of the supply of water from one side to the other, including, *inter alia*, reliability of supply, quality of supplied water, schedule of delivery and off-set of debts.

Mutual Cooperation

20. Both sides will cooperate in the field of water and sewage, including, *inter alia*:

   a. Cooperation in the framework of the Israeli-Palestinian Continuing Committee for Economic Cooperation, in accordance with the provisions of Article XI and Annex III of the Declaration of Principles.

   b. Cooperation concerning regional development programs, in accordance with the provisions of Article XI and Annex IV of the Declaration of Principles.

   c. Cooperation, within the framework of the joint Israeli-Palestinian-American Committee, on water production and development related projects agreed upon by the JWC.

   d. Cooperation in the promotion and development of other agreed water-related and sewage-related joint projects, in existing or future multi-lateral forums.

   e. Cooperation in water-related technology transfer, research and development, training, and setting of standards.

   f. Cooperation in the development of mechanisms for dealing with water-related and sewage related natural and man-made emergencies and extreme conditions.

   g. Cooperation in the exchange of available relevant water and sewage data, including:

      (1) Measurements and maps related to water resources and uses.

(2) Reports, plans, studies, researches and project documents related to water and sewage.

(3) Data concerning the existing extractions, utilization and estimated potential of the Eastern, North-Eastern and Western Aquifers (attached as Schedule 10).

Protection of Water Resources and Water and Sewage Systems

21. Each side shall take all necessary measures to prevent any harm, pollution, or deterioration of water quality of the water resources.

22. Each side shall take all necessary measures for the physical protection of the water and sewage systems in their respective areas.

23. Each side shall take all necessary measures to prevent any pollution or contamination of the water and sewage systems, including those of the other side.

24. Each side shall reimburse the other for any unauthorized use of or sabotage to water and sewage systems situated in the area under its responsibility which serve the other side.

The Gaza Strip

25. The existing agreements and arrangements between the sides concerning water resources and water and sewage systems in the Gaza Strip shall remain unchanged, as detailed in Schedule 11.

SCHEDULE 8
Joint Water Committee

Pursuant to Article 40, paragraph 15 of this Appendix, the obligations and responsibilities of the JWC shall include:

1. Coordinated management of the water resources as detailed hereunder, while maintaining the existing utilization from the aquifers as detailed in Schedule 10, and taking into consideration the quantities of additional water for the Palestinians as detailed in Article 40.

   It is understood that the above-mentioned Schedule 10 contains average annual quantities, which shall constitute the basis and guidelines for the operation and decisions of the JWC.

a. All licensing and drilling of new wells and the increase of extraction from any water source, by either side, shall require the prior approval of the JWC.

b. All development of water resources and systems, by either side, shall require the prior approval of the JWC.

c. Notwithstanding the provisions of a. and b. above, it is understood that the projects for additional water detailed in paragraph 7 of Article 40, are agreed in principle between the two sides. Accordingly, only the geohydrological and technical details and specifications of these projects shall be brought before the JWC for approval prior to the commencement of the final design and implementation process.

d. When conditions, such as climatological or hydrological variability, dictate a reduction or enable an increase in the extraction from a resource, the JWC shall determine the changes in the extractions and in the resultant supply. These changes will be allocated between the two sides by the JWC in accordance with methods and procedures determined by it.

e. The JWC shall prepare, within three months of the signing of this Agreement, a Schedule to be attached to this Agreement, of extraction quotas from the water resources, based on the existing licenses and permits. The JWC shall update this Schedule on a yearly basis and as otherwise required.

2. Coordinated management of water and sewage systems in the West Bank, as follows:

a. Existing water and sewage systems, which serve the Palestinian population solely, shall be operated and maintained by the Palestinian side solely, without interference or obstruction, in accordance with the provisions of Article 40.

b. Existing water and sewage systems serving Israelis, shall continue to be operated and maintained by the Israeli side solely, without interference or obstructions, in accordance with the provisions of Article 40.

c. The systems referred to in a. and b. above shall be defined on Maps to be agreed upon by the JWC within three months from the signing of this Agreement.

d. Plans for construction of new water and sewage systems or modification of existing systems require the prior approval of the JWC.

SCHEDULE 9
Supervision and Enforcement Mechanism

Pursuant to Article 40, Paragraph 17 of this Appendix:

1. Both sides shall establish, upon the signing of this Agreement, no less than five Joint Supervision and Enforcement Teams (JSETs) for the West Bank, under the control and supervision of the JWC, which shall commence operation immediately.

2. Each JSET shall be comprised of no less than two representatives from each side, each side in its own vehicle, unless otherwise agreed. The JWC may agree on changes in the number of JSETs and their structure.

3. Each side will pay its own costs, as required to carry out all tasks detailed in this Schedule. Common costs will be shared equally.

4. The JSETs shall operate, in the field, to monitor, supervise and enforce the implementation of Article 40 and this Schedule, and to rectify the situation whenever an infringement has been detected, concerning the following:

    a. Extraction from water resources in accordance with the decisions of the JWC, and the Schedule to be prepared by it in accordance with subparagraph 1.e of Schedule 8.

    b. Unauthorized connections to the supply systems and unauthorized water uses;

    c. Drilling of wells and development of new projects for water supply from all sources;

    d. Prevention of contamination and pollution of water resources and systems;

    e. Ensuring the execution of the instructions of the JWC on the operation of monitoring and measurement systems;

    f. Operation and maintenance of systems for collection, treatment, disposal and reuse, of domestic and industrial sewage, of urban and agricultural runoff, and of urban and agricultural drainage systems;

g. The electric and energy systems which provide power to all the above systems;

h. The Supervisory Control and Data Acquisition (SCADA) systems for all the above systems;

i. Water and sewage quality analyses carried out in approved laboratories, to ascertain that these laboratories operate according to accepted standards and practices, as agreed by the JWC. A list of the approved laboratories will be developed by the JWC;

j. Any other task, as instructed by the JWC.

5. Activities of the JSETs shall be in accordance with the following:

a. The JSETs shall be entitled, upon coordination with the relevant DCO, to free, unrestricted and secure access to all water and sewage facilities and systems, including those privately owned or operated, as required for the fulfillment of their function.

b. All members of the JSET shall be issued identification cards, in Arabic, Hebrew and English containing their full names and a photograph.

c. Each JSET will operate in accordance with a regular schedule of site visits, to wells, springs and other water sources, water works, and sewage systems, as developed by the JWC.

d. In addition, either side may require that a JSET visit a particular water or sewage facility or system, in order to ensure that no infringements have occurred. When such a requirement has been issued, the JSET shall visit the site in question as soon as possible, and no later than within 24 hours.

e. Upon arrival at a water or sewage facility or system, the JSET shall collect and record all relevant data, including photographs as required, and ascertain whether an infringement has occurred. In such cases, the JSET shall take all necessary measures to rectify it, and reinstate the status quo ante, in accordance with the provisions of this Agreement. If the JSET cannot agree on the actions to be taken, the matter will be referred immediately to the two Chairmen of the JWC for decision.

f.   The JSET shall be assisted by the DCOs and other security mechanisms established under this Agreement, to enable the JSET to implement its functions.

g.   The JSET shall report its findings and operations to the JWC, using forms which will be developed by the JWC.

SCHEDULE 10
Data Concerning Aquifers

Pursuant to Article 40, paragraph 20 and Schedule 8 paragraph 1 of this Appendix:

The existing extractions, utilization and estimated potential of the Eastern, North-Eastern, and Western Aquifers are as follows:

Eastern Aquifer:

- In the Jordan Valley, 40 mcm to Israeli users, from wells;
- 24 mcm to Palestinians, from wells;
- 30 mcm to Palestinians, from springs;
- 78 mcm remaining quantities to be developed from the Eastern Aquifer;

- Total = 172 mcm.

North-Eastern Aquifer:

- 103 mcm to Israeli users, from the Gilboa and Besian springs, including from wells;
- 25 mcm to Palestinian users around Jenin;
- 17 mcm to Palestinian users from East Nablus springs;

- Total = 145 mcm.

Western Aquifer:

- 340 mcm used within Israel;
- 20 mcm to Palestinians;
- 2 mcm to Palestinians, from springs near Nablus;

- Total = 362 mcm.

All figures are average annual estimates.

The total annual recharge is 679 mcm.

SCHEDULE 11
The Gaza Strip

Pursuant to Article 40, Paragraph 25:

1. All water and sewage (hereinafter referred to as "water") systems and resources in the Gaza Strip shall be operated, managed and developed (including drilling) by the Council, in a manner that shall prevent any harm to the water resources.

2. As an exception to paragraph 1., the existing water systems supplying water to the Settlements and the Military Installation Area, and the water systems and resources inside them shall continue to be operated and managed by Mekoroth Water Co.

3. All pumping from water resources in the Settlements and the Military Installation Area shall be in accordance with existing quantities of drinking water and agricultural water.

   Without derogating from the powers and responsibilities of the Council, the Council shall not adversely affect these quantities.

   Israel shall provide the Council with all data concerning the number of wells in the Settlements and the quantities and quality of the water pumped from each well, on a monthly basis.

4. Without derogating from the powers and responsibilities of the Council, the Council shall enable the supply of water to the Gush Katif settlement area and Kfar Darom settlement by Mekoroth, as well as the maintenance by Mekoroth of the water systems supplying these locations.

5. The Council shall pay Mekoroth for the cost of water supplied from Israel and for the real expenses incurred in supplying water to the Council.

6. All relations between the Council and Mekoroth shall be dealt with in a commercial agreement.

7. The Council shall take the necessary measures to ensure the protection of all water systems in the Gaza Strip.

8. The two sides shall establish a subcommittee to deal with all issues of mutual interest including the exchange of all relevant data to the management and operation of the water resources and systems and mutual prevention of harm to water resources.

9. The subcommittee shall agree upon its agenda and upon the procedures and manner of its meetings, and may invite experts or advisers as it sees fit.

APPENDIX
C
Effects of Water Use on Biodiversity in the Study Area

THE YARKON AND OTHER COASTAL RIVERS

The Yarkon River once was the largest perennial river flowing to the Mediterranean in the study area. Its biodiversity has been rich, including fish of recreational and commercial value. Though the river has been infected with schistosomiasis, it has been used for recreation (hiking, boating, and angling) by the inhabitants of the highest density urban center of the study area—Tel Aviv. Increasing discharges of urban and industrial pollution into the river at first encouraged an invasion of hyacinth floating plant cover, but eventually eradicated most of the biota (including the schistosomiasis vectors). Finally, the impoundment of Ein Afek Spring reduced the river to a sewage stream at its upper reaches, and to a marine tidal stream at its lower reaches. Rather than providing aesthetic and recreational opportunities, the river ecosystem generated unpleasant smells and mosquito outbreaks.

The Taninim River has an annual flow of 50 million m³/yr. Half of this flow is provided by a large mountain watershed to the east of the river, and the other half by Timsah springs in the foothills of the watershed. Nature reserves along the course of the coastal section conserve the river's biodiversity, and they attract some quarter of a million visitors each year. However, water quantity and quality still steadily decline. Pumping from the aquifer feeding Timsah springs reduces its discharge. In flows into the river include untreated sewage of local townships, irrigation drainage water enriched with fertilizers and pesticides from agricultural fields and drainage aquaculture ponds. In addition, prior to reaching the nature reserve, half of the flow is diverted to feed fish ponds. As a result, the river is stressed by pollutants and increasing salinity, especially toward the end of summer.

The high concentration of pollutants and the low and slow flow promotes the spread of duckweed—a distinct signal of severe ecosystem change, and many aquatic species typical to this river have already disappeared. Yet, the Taninim River, with protected areas along its course, is the only perennially flowing coastal river in Israel because the natural low salinity of the Timsah springs, 1,200 mg Cl/l, is suitable for aquaculture but not for agriculture. However, the economic feasibility of desalinating this relatively low salinity water makes the 25 million m^3/y discharge of Timsah springs attractive for closing the water supply "gap" of exactly the same amount, forecasted for the city of Haifa by year 2000. If this project of impounding, desalinating, and transporting all the Timsah discharge is to be implemented, it is estimated that the Taninim River's current flow of 50 million m^3/yr will be reduced to 18 million m^3/yr of highly polluted water. This project will obstruct the outlet of the river to the sea, such that estuarine biodiversity will disappear (Ben-David, 1987). Thus, the Taninim River may become a case in which implementing desalination technology for domestic water supplies will kill the only functioning coastal river ecosystem west of the Jordan River.

THE JORDAN RIVER BASIN

The basin elevation ranges from 90 m above sea level to 400 m below sea level and includes three source streams, which create the northern section of the Jordan River. Most important of these three headwaters is River Dan, the only Israeli river that has a seasonally stable output. It has also a stable temperature, a year-round high oxygen saturation and a high number of species (156 aquatic animal species). The three streams cross the Hula Project region as a canal, then descend in the Jordan River's natural course to Lake Kinneret.[1] The Kinneret drains to the lower Jordan River, which discharges to the Dead Sea, which is a dead-end lake.

The major water management activities in the basin are the drainage of the Hula wetland and its subsequent management (See Box 4.2), and the management of Lake Kinneret, which is the major surface water storage of the State of Israel. The Hula project affected and is still affecting Lake Kinneret's water quality, the management of Lake Kinneret affects

---

[1]Lake Kinneret is also named the Sea of Galilee and Lake Tiberias.

the Lower Jordan River ecosystem, and both affect the economy, environment and biodiversity of the Dead Sea region. Each section of the Jordan
River basin is described below, from north to south.

THE HULA WETLAND

The Hula was a large wetland at the north end of Lake Kinneret. It is
now reduced in size by drainage and most of the wetland has been replaced by cropland. A nature reserve was reconstructed on a part of the
drained wetland and a section of the newly created cropland area is
flooded seasonally, until recently part of that section was intentionally
flooded. These vicissitudes had dramatic effects on the species richness
and composition of the Hula region. Dimentmann et al. (1992) showed
that 585 (612, if doubted or insufficiently described records are included)
aquatic animal species, excluding unicellular and parasitic species, were
recorded in the wetland prior to drainage. Of these, 19 were represented
by peripheral populations (for 14 and 5 species the Hula constituted the
southern and northern limit of the species' global distributions, respectively), and 12 were endemic to the Hula wetland (6 beetles, 2 dragonflies,
a flatworm, a fly, a frog, and a fish). The reconstructed nature reserve
lacks, 119 (20 percent) of the native species, including 11 of the 19 species
represented in the Hula by peripheral populations, and 7 of the 12 Hula
endemic ones. Seven species, among them a frog and a fish, have become
globally extinct. Furthermore, since the drainage of the Hula, 36 of the
species lost to the Hula have not been recorded anywhere else in Israel.
The birds are the best known group of the Hula, and the information
about this group is highly reliable. Of the 36 species breeding prior to
drainage, 10 ceased to breed after the drainage, but 5 were replaced by
species that had not bred there prior to drainage.

To summarize, the drainage of the Hula, a wetland that is relatively
small in global terms, resulted in a local loss of 119 species (plus 10 birds
species that ceased to breed there), the national loss of 36 species, and the
global loss of 7 animal species. On the other hand, 212 aquatic animal
species new to the Hula have been recorded after the drainage. Some of
these might have existed in the Hula prior to the drainage but escaped
attention. However, most of them are probably new colonizers, indicative of the changes in habitat extent and diversity, and in the quality of the
water, following the drainage and subsequent reconstruction efforts.

The Hula formerly supported a unique community of species. From
the north (Europe), west (Mediterranean basin), east (Iraq, Iran), and south
(Egypt, tropical Africa) of the Hula many species assembled into a unique
community in the Hula. Thus, although most of the species also exist
elsewhere, their combination, and hence their interactions, existed no-

where else. The ability of northern and tropical species to live together in the Hula developed from the high diversity of aquatic habitats in the Hula and the year-round discharge of springs with year-round stable moderate temperatures, thus providing the Hula with a refuge from extreme high summer and low winter water temperatures. Species belonging to each of these biogeographic categories were lost by the drainage, and hence, unique species interactions, probably related to unique ecosystem functions, were lost. Also, some dramatic natural phenomena, such as the upstream spawning migration into the Dishon stream of Lake Hula's three cyprinid species, are forever lost, though the species themselves have not become extinct.

The management of the Hula wetland is a clear case of water resource development raising conflicts between agriculture and the ecosystem services provided to society by biodiversity. Agricultural development in the reclaimed Hula lands potentially conflicts with water quality downstream in Lake Kinneret.

There is also local conflict between the economic benefit to the farmers now cropping the Hula land, and the economic benefit of the "aesthetic services" provided by a recreation in the Hula wetland. A similar though not identical conflict has developed in the other large wetland of the region. The Azraq oasis in Jordan drew international attention in the 1960s (Mountford, 1965; Nelson, 1973) as a hotspot of desert biodiversity, with cultural and aesthetic values. Recently, however, the functioning of this oasis as a wetland declined, due to exploitation of its water source to supply the increasing demands of the urban population of Jordan, notably that of Amman. The oasis of Azraq now has lost most of its aesthetic value.

LAKE KINNERET/LAKE TIBERIAS/SEA OF GALILEE

The structure and function of the Lake Kinneret ecosystem, described by Gophen (1995), are relevant to the ways in which the lake's biodiversity and water quality are affected by the management of the watershed and the water level of the lake itself.

The Effect of the Watershed

Lake Kinneret stores between 3,903 and 4,301 million m^3 of water, depending on the water level, and the mean annual natural in flow is 940 million m^3. The lake's water turn over on average once in every 4.4 years, a relatively short residence time for lakes. The area of Kinneret's watershed is 2,730 km^2, and the ratio of this area to the lake's volume is 0.68, a relatively high value for lakes. The two indices mean that, first, the watershed, mostly agricultural land, introduces high quantities of fertilizers and other pollutants relative to the lake's volume; and second, that water use prior to discharge to the lake and the use of the lake's water jointly reduce the lake's volume and shorten the turnover time of the lake's water, thus increasing the salinity and the nutrient contents of the lake. The Jordan River, which drains large parts of the watershed, annually discharges to the lake 1,610 tons of nitrogenous compounds and 130 tons of phosphorous compounds annually. Much of the nitrogen is denitrified in the Kinneret and released to the atmosphere. Much of the phosphorus discharge is deposited at the lake bottom in an inert form, not available to algae.

Effects of Managing the Lake Water Levels

The lake is the major operational reservoir for water supply to Israel. Ecosystem functions of the lake are involved in determining the quality of its water for domestic and agricultural use. A high density of microalgae and high salinity reduce the quality of the lake as a provider of potable water and irrigation water, respectively. Microalgae cause undesirable odor and taste and secrete toxic compounds and materials that interfere with the disinfecting process of potable water. The density of microalgae is regulated by various factors, among them grazing microcrustaceans, which in turn are controlled by predatory fish. Salinity derives from discharge of saline springs at the bottom and the edges of the lake and from the discharge of the Jordan River. The volume of the lake, expressed by its water level, affects the two water quality attributes, microalgae and salinity. With respect to salinity, hydrostatic pressure may control the saline discharge from the springs at the bottom of the lake. The microalgae are limited at times by mineral resources, mainly dissolved phosphates. The lake's bottom is rich in particulate phosphorous compounds, and their dissolution, hence their availability as a resource for microalgal growth, is controlled by the concentration of CO_2 in the lower water strata (the hypolimnion). Prior to human intervention, the water level was determined by variations in inflows (rainfall, storm runoff, river discharge) and outflows (evaporation and the outlet discharge to the lower Jordan River). The level fluctuated within a range of 1.3 m with the high bound determined by winter floods and the low bound by the depth of the bottom of the outlet to the Jordan River.

The management of the lake involves manipulations of both inflow and outflow. Extraction from sources (impoundments of springs discharging to the Jordan River, damming of storm water in the watershed) affect the inflow. Pumping into the National Carrier, damming the outlet to the lower Jordan River and manipulating the dam all affect the outflow. Legislation currently determines the upper and lower bounds of the water level and prevents damage to shore installations (pumps, piers, recreation facilities) and their operation. The motivation for lowering the level as much as possible is the need to free space for storage of the winter inflow, even if this is at the expense of "losing" water to the Dead Sea. Currently the fluctuation range of the managed lake is 4.1 m and the mean water level is 0.5-1 m lower than the natural mean level of the lake.

These managed low levels may lower water quality. With respect to salinity, there is a risk of increased discharge of the saline springs at the bottom of the lake. Countering this risk is the "salty carrier." This is a diversion which captures saline springwater flowing in the Kinneret and transports it directly to the lower Jordan River, thus bypassing the lake and freeing it from a third of its annual salt input.

Low water levels, which reduce the volume of water in the lake, affect the biota of the lake by changing the concentrations of dissolved chemicals. Microalgae increase in response, and at first are controlled by planktonic small crustaceans. But soon the numbers of crustaceans and reduced fish increase their predation on the crustaceans, thus releasing the microalgae from their control. Fertilizers transported from the watershed also enrich the lake with nutrients, contributing to the increase of microalgae. Low water levels also affect the littoral and shallow estuaries (Gasith and Gafny, 1990). The turbidity associated with low water levels deposits muddy sediments on pebbles and other hard substrates, thus reducing their quality for fish reproduction. Abrupt fluctuations of the water level also reduce the quality of stones for egg attachment (due to development of microalgal mat). Altogether, the lowering of the lake level increased the proportion of sandy bottom of the littoral from 10 percent to 60 percent, thus reducing the extent of pebbles required for fish reproduction. Lengthy exposures encourages colonization of riparian and terrestrial vegetation, and reflooding of these areas in years of high runoff increases the organic load of the lake, through the death and decomposition of this vegetation. Low levels add to the management problems for wetland nature reserves along the northern and eastern coasts.

Effects of Further Lowering the Lake's Water Level

Increasing water level fluctuations from the natural 1-2 m to 4 m during 1969 to 1993 did not cause apparent deterioration in the functioning of the lake's ecosystem. This observation has been interpreted as indicating the high resilience of this ecosystem and has encouraged considering further lowering of the Lake Kinneret level. However, as of 1994, hitherto unknown blooms of nitrogen-fixing cyanobacteria (blue-green algae) have occurred. Besides indicating a reduction in the quality of the water, and the fact that these blooming algae are known to produce chemicals toxic to aquatic animals and humans, this change indicates that the Kinneret ecosystem has an unpredictable potential for change. Further lowering of the level may generate further instability in the structure and function of the Kinneret ecosystem.

Besides increasing the risk of urban wastewater pollution of pumped water (since by lowering the level, pumping points will become closer to coastal pollution sources), further lowering of the water by one meter may result in the following ecosystem effects (Zohary and Hambright, 1995): (1) reduced water quality due to increased suspended materials and algal productivity brought about by further decrease in the summer volume of the lake's lower strata, hence faster rate of prevalence of anaerobic conditions, earlier accumulation of sulfides, earlier release of phosphate from the lake's bottom, and additional 33 percent increase in dissolved phosphate; and (2) reduced ecosystem stability, via food chain fluctuations brought about by changes in the food chain of littoral dependent fish-zooplankton-phytoplankton.

THE LOWER JORDAN RIVER AND THE DEAD SEA

The "Salty Carrier," which discharges the coastal saline springs' water to the lower Jordan River, significantly reduces the salinity of the lake's water, thus counteracting the salinization trend associated with lowering the level of the lake. But this arrangement, as well as the overall reduction of the inflow from the Kinneret to the lower Jordan River, increases the salinity of the river to the point of losing significant components of its aquatic and riparian biodiversity and changing the structure of its biotic community and ecosystem functions. The discharge of Lake Kinneret water through the lower Jordan River to the Dead Sea is a "loss" to the water budget of the region, minimized by the prescribed low levels of Lake Kinneret. But this is not necessarily the only loss.

The reduced discharge lowers the level of the Dead Sea, and this low level and the resulting receding coastline negatively affect Dead Sea coastal installations—industrial, mining, health (spas), and recreational. Also, the low level and receding coastline of the Dead Sea affect the freshwater oases along the coast: Ein Fashkha and Ein Tureiba on the western coast are wetland nature reserves used both for recreation and for conservation of a unique biodiversity. They suffer from management problems due to constant changes in their water budget. Furthermore, not many plants are able to colonize the large expanses of the exposed, highly saline former littoral now surrounding the Dead Sea. There is a potential risk of an "Aral Sea effect" developing there, in which wind-blown minerals of the exposed lake surface affect biodiversity at a distance from the coast.

REFERENCES

Ben-David, Z. 1987. Taninim River —nearly the end of the road. Society for the Protection of Nature in Israel, report (in Hebrew).

Dimentman, Ch, H. J. Bromley, and F. D. Por. 1992. Lake Hula. Reconstruction the fauna and hydrobiology of a lost lake. Jerusalem: Israel Academy of Sciences and Humanities.

Gasith, A., and S. Gafny. 1990. Effect of water level fluctuation on the structure and function of the littoral zone. In Large Lakes, Ecological Structure and Function, M. M. Tilzer and C. Serruya, eds. New York, N.Y.: Springer-Verlag.

Gophen, M. 1995. Whole lake biomanipulation experience: Case study of Lake Kinneret (Israel). In Lake/Reservoir Management by Food Chain Manipulation, R. De Bernardi and G. Giussani, eds. ILEC-UNEP 7,171-184.

Mountford, G. 1965. Portrait of a Desert. London, U.K.: Collins.

Nelson, J. B. 1973. Azraq: Desert Oasis. London, U.K.: Allen Lane.

Zohary, T., and K. D. Hambright. 1995. State of knowledge relevant to effects of further lowering of the Kinneret's minimal level. Kinneret News 18,2-6 (in Hebrew).

Guidelines for Rehabilitation of Rivers

Friedler and Juanico (1996) proposed guidelines for water allocation for rehabilitation of rivers in Israel. This appendix therefore focuses on Israel's rivers but the biophysical principles apply to all rivers in the study area, including the Jordan River. The river should have waters that sustain biodiversity, provide the ecosystem service of "open space," and also allow for economic development along the river course. With respect to water quantity, at less than 10 percent of natural base flow (excluding storm water) the river ceases to function as a river, whereas this 10 percent flow quantity can be tolerated, provided this low flow only occurs for short periods. To maintain the aquatic and riparian biodiversity, however, 30 percent of natural base flow is minimal sustained average. Flood flows are critical too. Floods remove deposits accumulated during the long summer that would otherwise obstruct the flow. Water quality standards are proposed for chlorination (carried out to meet health requirements, but which can be toxic to all fish), for organic load (high load causes dangerous anoxia), for ammonium (which can generate toxic ammonia) and for pH and salinity (which must be kept within ranges prevalent in the natural stream). A prerequisite for meeting these standards is that the velocity should not be lower than 0.2 m/sec (with water width and depth of at least 5 m and 0.5 m, respectively). It should be noted that compliance with human health regulations alone can still be destructive to the river's aquatic biodiversity. The recipe and timetable for the rehabilitation of the "dead" rivers of Israel is as follows: (1) a legal procedure of water allocation to the river should be completed, and the discharge of the allocated water should proceed; (2) point pollution sources along the river course should be removed, and at the same time nonpoint pollution sources should be identified and controlled; (3) measures for secondary treatment of sewage sources should be taken; and (4) tertiary treatment should be applied, and facilities for pooling wastewater to control the flow of the river should be constructed and operated.

REHABILITATION OF THE YARKON, ALEXANDER, AND SOREQ RIVERS

The 28 km of the Yarkon River meander through Israel's most densely populated area. The discharge of the source of Yarkon river, Ein Afek springs, was 220 to 200 million m^3/yr of water prior to the transportation of most of it to irrigate the Negev in 1955. The river died out, and to rehabilitate it, 65 million m^3/yr of water were allocated in 1992. The effect on the river was most dramatically expressed in the return of its fish fauna. But later this amount was reduced, from the pressures of other users, and the state of the river deteriorated. The master plan for rehabilitating the river (Rahamimov, 1996), commissioned by the Yarkon River Authority, which was established in 1988, paves the way for full rehabilitation of the river. The master plan follows the guidelines for river rehabilitation in Israel and is based on the premise that only 9 million m^3/yr of freshwater could be allocated, with the rest replaced by treated wastewater, 12 million m^3/yr of which is already allocated. The allocation to be released from the impounded Ein Afek springs, together with the allocation of treated wastewater, is to guarantee 10 percent of the original flow—2,500 m^3/hr. The water will be sold to users along the river course, that is to authorities who will operate parts of the riparian areas as recreational areas. Finally, prior to reaching the last, saline section of the river, the water will be impounded for conventional use. Thus, except for the little water lost by evaporation, there will be no losses to the national water budget. Percolated water will recharge the aquifer, and the rest will be sold twice. This arrangement should fully compensate for the cost of impounding the water downstream rather than upstream near the source (i.e., will cover the cost of uplifting the water for users above the point of impoundment). Hundreds of tons of garbage have been removed from the river to restore its original depth, the river's banks have been cleaned up, reinforced, and raised, sewage treatment plants in some cities discharging wastewater to the river have been inaugurated, and mosquito larvae are controlled by introduced predatory fish (*Gambusia*) and by seasonal application of *Bacillus thuringiensis israelensis* (BTI), a mosquito larvae—specific pathogen, inert to all other forms of life. The last approach is a demonstration of the potential use of local biodiversity—this pathogen, discovered in an Israeli ephemeral pond, has become the major means of control of local mosquitoes, and also a source of income as an export product.

Following the example of the Yarkon River, the 44 km of the Alexander river, inhabited by a relatively large and sole population of the Nile softshell turtle (*Trionyx triunguis*), is being rehabilitated as of 1995, and now receives 225,000 m^3/hr from various sources. The Soreq river used to be a permanent coastal river, with winter floods added by stream from the Soreq mountains. This stream currently is a permanent flow of Jerusalem's 14 million m^3/yr of sewage, 20 percent of which is lost by percolation prior to reaching the coastal plain. The remainder reaches treatment plants in the foothills and is used for irrigating cotton. A proposal has been made to create a continuous flow along the coastal part, with 15 million m^3/yr of treated wastewater from local towns.

SECURING ALLOCATIONS OF WATER FOR AQUATIC AND RIPARIAN ECOSYSTEMS

The Legal Status of Water in Israel and the Quota of Water for Nature

All the waters of Israel belongs to the State, but legislation concerning pricing distinguishes between pumped water and natural water, such as direct rainfall, surface runoff, natural, open streams, pools, and other water bodies. All protected areas in Israel, both nature reserves and national parks, are the property of government authorities, the Nature Reserves Authority and National Parks Authority (recently united into one authority reporting to the Ministry of the Environment). Water sources within protected areas, as elsewhere, are managed by the Water Commissioner. However, natural water that is charged to users elsewhere is not charged to the Nature Reserves Authority, who is the user. Nevertheless, in each specific case permits are negotiated between the Nature Reserves Authority and the Water Commissioner, but other traditional and potential users who contest the proposed water use are involved in the negotiations. Legally, all agreed-upon permits are temporary and can be revoked at any time by the Water Commissioner, with no provision for alternative sources of water.

Criteria for Allocations of Water for Nature

Research has not been carried out in the study area to determine the quantity and quality of water required by natural ecosystems for maintaining their biodiversity and providing their services. Guidelines are also lacking for drylands outside the region. The rule of thumb of the Israeli Nature Reserves Authority is to negotiate for the natural quantity and quality first, and finally to settle on the best compromise possible. Once an allocation is determined, the Nature Reserves Authority guards it tenaciously against all future attempts to challenge this allocation. Such attempts become frequent, as water demands in the study area increase. It is likely that the agreed allocations will become unacceptable to alternative users, and the authorities will then face pressures to reduce the allocations for nature. Just as for agriculture, natural ecosystems will have to compete with agriculture for treated wastewater. Precise objectives will be needed for each aquatic, riparian, and other water-dependent site in the study area (referring, e.g., to the type of biodiversity to be maintained and the types of ecosystem services required), and studies will be needed to determine the minimal required allocation of quantity and quality. Indicators, benchmarks, and monitoring programs for each of the sites will have to be identified and carried out for reviewing and updating water allocations.

The Case of Allocation for the Ein-Geddi Reserve

Several year-round discharging springs nurture the Ein-Geddi oasis near the Israeli coast of the Dead Sea. The biodiversity of this oasis is an Ethiopian relict island in an "ocean" of Saharo-Arabian biota. Significant species for ecosystemic function as well as tangible inspirational value, are the ibex and its predator, the leopard. The oasis is a nature reserve, but Kibbutz Ein-Geddi, located next to the oasis, is provided with water by the Ein-Geddi springs for its agricultural and domestic needs. Legally, the Kibbutz is entitled to use all the discharge of the spring, which are fully impounded. However, the Kibbutz uses the spring for potable domestic use only. The remainder flows in a stream that feeds the reserve, but is then transported for other domestic uses of the Kibbutz and for its agriculture. Only what remains then, which fluctuates much between years, is to be left solely for the reserve. So far the overall Kibbutz needs hae been less than the discharge, and therefore there have always been some water in the reserve. Yet agreements have been negotiated, to guarantee 20 m^3/hr for the reserve. In practice, of the 339 m^3/hr average discharge of all Ein Geddi springs, 264 m^3/hr is for all users, and 75 m^3/hr are left for biodiversity.

It has now been realized that to prevent total extinction of the reserve's water-dependent vegetation, 84 m^3/hr are required. The average annual gap of 9 m^3/hr has to be filled by increasing the allocation to the reserve at the expense of the upstream users, or by developing new water resources for nature, such as wells dug into the alluvial fans next to the

Dead Sea shores, which store flood and drainage water (Hendelsman, 1990).

Current Allocations in Israel

Negotiations on water quota in Israel have so far yielded an array of interesting arrangements for allocating water to nature (i.e., for promoting biodiversity and ecosystem services) and sharing water sources with other users. The following is a review of these arrangements, identifying the amounts of water involved.

Springs and Small Streams. Water allocation arrangements for springs and small streams include the following:

- One stream with a rare aquatic plant and another stream with an endemic Israeli fish are allocated the whole discharge of their respective springs; in two nature reserves, the flow is impounded after leaving the protected areas.
- Five nature reserves are allocated a fixed quantity of the discharge, ranging from 10,000 m^3/yr in one reserve to 1,051,200 m^3/yr in another.
- Three nature reserves receive variable allocations, depending on the variations in the discharge: one reserve is allocated a minimum of 60 m^3/hr, but when the discharge is 120-200 m^3/hr, half of the discharge is released for nature, and when it is above 200 m^3/hr, 100 m^3/hr is allocated to nature; in another reserve, 1 m^3/sec is allocated in an average year, but in dry years allocation is curtailed proportionally to the curtailment of allocation to other users, provided that at least 0.5 m^3/sec is allocated for nature; and in the third reserve, the allocation is not less than 52,560 m^3/yr, with extra amounts depending on other users.
- For three reserves, there are priority rights and allocations set for other users, and nature receives the remainder, if any;
- In two reserves, all water can be taken for other uses, provided water flow persists even in dry summer, at a continuous flow in the channel.
- A mixed arrangement has been made for Har Meron Nature Reserve, the largest nonaquatic protected area in the Mediterranean highlands: all springs within the reserve except one are allocated for nature; for the other, nature and other users share the discharge equally.
- An interesting case is that of Gush Halav Spring near the Har Meron reserve. This spring traditionally served the needs of the village of Gush Halav for generations. With recent increasing water demands the village was connected to the national grid and the domestic and agricultural use of the spring ceased, its water becoming the property of nature

by default. Recently, an entrepreneur requested use of the spring water for a mineral water industry. The negotiations involved the Nature Reserves Authority who claimed that all the 7 m³/hr and 10 m³/hr in summer and winter respectively are required for nature, and therefore objected to a permit for any industrial use.

Rivers. In mountain river reserves, sources are impounded, but some water is later released in several points along the course of the river. In Snir stream there are two points for releasing 11,388 m³ and 14,016 m³ respectively in an average year and 6,132 m³ and 7,446 m³ respectively in a dry year. In Kziv stream there are three points of release of fixed quantities irrespective of the type of year, totaling 328,000 m³/yr. The Taninim River is the only coastal river that has a nature reserve status. All other coastal rivers of Israel have died; some are at various stages of rehabilitation. Currently the river is allocated a minimum of 3,504,000 m³/yr. This is provided by a flow of 110 l/sec. When the flow is above 530 l/sec, on each 30 l/sec increment, there is a 10l/sec increment in the allocation. When the flow is smaller than 420 l/sec, on each 30 l/sec depletion, the allocation is reduced by a 10 l/sec.

Wetlands. The Hula nature reserve is fed by a spring, and by a canal that replaces in part the predrainage discharge of the Jordan River. The spring water is divided between agriculture and the reserve, such that when the discharge is below 300 m³/hr all this amount is allocated to the reserve; when the discharge is lower than 450 m³/hr, 600 m³/hr, and 750 m³/hr, the allocation to agriculture is 100 m³/hr, 125 m³/hr, and 150 m³/hr, respectively; and when it is more than 750 m³/hr, agriculture receives 200 m³/hr.

   In effect, in winter the allocation is 1,000 m³/hr for agriculture and 500 m³/hr, as well as the excess of agriculture for the reserve. Altogether the average allocation is 1,532,000 m³/yr for the reserve. The Western Canal discharges 4 million m³/yr to the reserve, but most of this is marginal water—wastewater and aquaculture effluents. In a much smaller wetland in the Hula region, the Gonen meadow, the water allocation is small—it maintains moisture in the reserve.

Allocation of "Nonnatural" Water for Nature

   There are two cases of specific allocation of "nonnatural" water to nature: in one case an overflow of storage next to a reserve is allocated to the reserve. In another case the reserve is allocated 10,000 m³/yr from a well. Nonnatural water must be paid for by the Nature Reserves Authority, and so these allocations seem to have been not used ever.

The overall amount used in Israel for natural springs and small streams in nature reserves, which is contested by other users, currently is a minimum allocation of 18,200,760 m^3/yr. This amount can be increased to at least 32,412,000 m^3/yr. The maximum allocation is hard to calculate, since it requires data on the between-year variability in discharges of the various sources, as well as the variability in the allocations to other users. Large streams, rivers, and wetlands that are nature reserves are allocated at least 5,378,578 m^3/yr. The maximum allocation, again, can not be calculated easily.

Thus, altogether the nature reserves of Israel are allocated between 23,579,338 m^3/yr and 56,016,742 m^3/yr, and more water is allocated depending on between-year variabilities in discharges and other uses. It seems, though, that legislation does not always guarantee the allocation, either because discharges are lower than expected, or because other users manage to obtain more than their share. For example, the allocations for three reserves—Hula, Ein Afek, and Kziv stream—combined for each reserve for the years 1991, 1992, and 1993 in thousands m^3/yr were 5,724, 5,624 and 5,624, respectively, but the actual use by the reserves was only 3,682, 3,162, and 4,037, respectively. Nature reserves and rivers are also allocated water of lower quality (wastewater, untreated and treated), but reliable data on quantities are not yet available. Altogether, the 24 to 56 million m^3/yr legally allocated for biodiversity and ecosystem services in Israel accounts for 0.9 percent to 2.0 percent of the total renewable water resources west of the Jordan Rift Valley (see Table 2.2). It should be borne in mind that, though much of this water evaporates, a substantial part recharges the aquifers and thus becomes available for further, possibly alternative, uses.

ENVIRONMENTAL IMPACT ASSESSMENTS

Israeli legislation requires environmental impact assessments for major development projects. Whether or not a project is major is decided by the Ministry of the Environment. However, many developers perform environmental surveys at their own initiative, so that they do not invest in projects that are likely to be rejected later on environmental grounds. The major water projects carried out in Israel have not been explored for their environmental effects prior to their execution. But this is gradually changing. Projects for pumped energy, such as creating artificial reservoirs above the Lake Kinneret/Lake Tiberias/Sea of Galilee or managing part of the Jordan River channel, have undergone a thorough environmental assessment process. Plans for impounding floodwater in the Negev are currently undergoing environmental assessment.

Most of these environmental impact assessments, however, concentrate on effects on human health, aesthetics, risks of air and water pollution, and the risk for endangered species. The recognition of the significance of ecosystem services and the role of biodiversity in providing these services, beyond the endangered "flagship" species, is not yet expressed in environmental impact assessments.

The major future undertaking is to assess the impacts of wastewater use, for agriculture and for biodiversity, on biodiversity and ecosystem services. With respect to using treated wastewater to sustain biodiversity, predictions for water demand and use suggest that the current allocations for nature, as well as the recommendations of environmental impact assessments may not be respected in the future. Therefore, just as treated wastewater may in the future replace freshwater for agriculture, wastewater allocation for maintaining aquatic ecosystems and water-dependent terrestrial ecosystems may replace the current allocation of freshwater. Research is therefore required now for determining the effects and the technologies appropriate for this future substitution.

Urban development in one of the prime agricultural areas of Israel, the coastal plain, is currently motivating the transfer of agriculture to the northern Negev, namely, to the semiarid belt of Israel, which is also the climatic transition zone between desert and nondesert in Israel. The water required to sustain this agricultural development will be the effluents of the urban areas in the coastal plain. The treated wastewater is already transported from there to the northern Negev, and volume of water transportation will increase and proliferate. This new agricultural development will replace natural ecosystems whose biodiversity is of prime significance (Safriel et al., 1994), and which have not undergone any process of environmental assessment. It is certainly necessary to assess the impact of this development, but it is also essential to explore methods of regional planning that will minimize the damage especially to habitats that harbor indispensable biogenetic resources.

REFERENCES

Friedler, E., and M. Juanico. 1996. Allocation of water for rehabilitation of selected rivers in Israel. Report to KKL.

Hendelsman, E. 1990. Water as a source of life. Nature Reserves Authority (in Hebrew).

Rahamimov, A. 1996. Master Plan for the Yarkon River. Tel-Aviv, Israel: Yarkon River Authority (in Hebrew).

Safriel, U. N., S. Volis, and S. Kark. 1994. Core and peripheral populations and global climate change. Israel J. of Plant Sciences 42:331-345.

APPENDIX
E

Bibliography

Abu Taleb, M. F., J. P. Deason, and E. Salameh. 1991. The Jordan River Basin. Washington, D.C.: The World Bank.

Abu-Hijleb, L. 1993. Investigation of Potential Applications of Rainwater Catchment Systems in the Gaza Strip and an Exploration of Appropriate Sanitation Systems. Report for the International Water Engineering Center, University of Ottawa.

Ahiram, E. and H. Siniora. 1994. The Gaza Strip Water Problem: An Emergency Solution for the Palestinian Population. Pp. 261-271 in Water and Peace in the Middle East. Proceedings of the First Israeli-Palestinian International Academic Conference on Water, J. Isaac and H. Shuval, eds., Zurich, Switzerland, December 10-13, 1992. New York, N.Y.: Elsevier.

Al-Kharabsheh, A., R. Al-Weshah, and M. Shatanawi. 1997. Artificial Groundwater Recharge in the Azraq Basin (Jordan), Dirasat. Agricultural Sciences 24(3)September.

Al-Weshah, R. A. 1992. Jordan's Water Resources: Technical Perspective. Water International 17(3)124-132.

Amiran, D. H. K. 1995. Rainfall and Water Management in Semi-Arid Climates: Israel as an Example. Research Report No. 18. Jerusalem Institute for Israel Studies.

Anderson, E. W. 1990. Water Geopolitics in the Middle East: The Key Countries. Washington, D.C.: Center for Strategic and International Studies.

Arlosoroff, S. 1995. Promoting water resource management in the Middle East. Int'l. Water Irrig. Rev. 15:6-16.

Assaf, D. 1996. From Stones to Structures: A Sustainable Future for Development in the West Bank—Palestine. Dissertation. Seattle: University of Washington.

Assaf, K., N. Al-Khatib, E. Kally, and H. Shuval. 1993. A Proposal for the Development of a Regional Water Master Plan. Israel/Palestine Center for Research and Information, October.

Awerbuch, L. 1988. Desalination Technology: An Overview. Chapter 4. Pp. 53-64 in The Politics of Scarcity: Water in the Middle East, J. R. Starr, and D. C. Stoll, eds. Boulder, CO: Westview.

Baskin, G., ed. 1993. Water: Conflict or cooperation. Israel/Palestine Issues in Conflict Issues for Cooperation. Israel/Palestine Center for Research and Information II(2), March.

Benvenisti, E., and H. Gvirtman. 1993. Harnessing international law to determine Israeli-Palestinian water rights: The mountain aquifer. Natural Resource J. 33(3):543-568.

Belheisi, M. 1992. Jordan's water resources and the expected demand through the years 2000 and 2010, detailed. In Jordan's Water Resources and Their Future Potential, A. Garber, and E. Salameh, eds. Amman: Freidrich Ebert Stiftung.

Benblidia, M., J. Margar, and D. Vallee (under the direction of B. Glass). Water in the Mediterranean Region: Situations, Perspectives and Strategies for Sustainable Water Resources Management. Blue Plan Regional Activity Centre, Sophia Antipolis, France.

Berkoff, J. 1994. A Strategy for Managing Water in the Middle East and North Africa. Washington, D.C.: The World Bank.

Beschorner, N. 1992. Water and instability in the Middle East. London, U.K.: International Institute for Strategic Studies. Adelphi: Paper 273, Winter.

Bingham, G., A. Wolf, and T. Wohlgenant. 1994. Resolving Water Disputes: Conflict and Cooperation in the United States, the Near East, and Asia. Arlington, VA: ISPAN for USAID.

Biswas, A. K., J. Kolaro, M. Morahami, J. Waterbury, and A. Wolf. 1997. Core and Periphery: A Comprehensive Approach to Middle Eastern Water. Middle East Water Commission. Delhi: Oxford University Press.

Bitton, G., B. L. Damron, G. T. Edds, and J. M. Davidson. Sludge-Health Risks of Land Application. Stoneham, Mass.: Ann Arbor Science/Butterworths.

Bjorklund, G. 1992. Comprehensive Assessment of the Freshwater Resources of the World. Stockholm Environment Institute, Stockholm, Sweden. 33 pp.

Brown Weiss, E. 1989. In Fairness to Future Generations. New York and Tokyo: Transnational and United Nations University (also in French and Japanese).

Brown Weiss, E. 1995. Intergenerational Fairness for Water Resources. Environmental Law and Policy 25:231-236.

Caponera, D. A. 1993. Legal aspects of transboundary river basins in the Middle East: The Al Asi (Orontes), the Jordan and the Nile. Natural Resources J. 33(3):629-664.

Casageldin, I. 1995. Water Resources Management: A New Policy for a Sustainable Future. Water International 20:15-21.

City of Northglenn Colorado's Wastewater Treatment Plant. 1984-1992. Annual Reports (1984-1992) showing Treatment Effectiveness of Three-Cell Aerated/Storage Reservoir. City of Northglenn, Colorado Wastewater Treatment Plant.

Downey, T. J., and B. Mitchell. 1993. Middle East water: Acute or chronic problem? Water International 18:1-4.

Duna, C. 1988. Turkey's Peace Pipeline. Chapter 7, pp. 119-124 in J. R. Starr, and D. C. Stoll, eds. The Politics of Scarcity: Water in the Middle East. Boulder, CO: Westview.

Eckstein, Z., D. Zakai, Y. Nachtom, and G. Fishelson. 1994. The Allocation of Water Sources Between Israel, the West Bank and Gaza: An Economic Viewpoint. Tel Aviv, Israel: Tel Aviv University.

Elmusa, S. 1993. Dividing the common Palestinian-Israeli waters: An international water law approach. Palestinian Studies XXII(3):57-77.

Elmusa, S. 1994. Towards an Equitable Distribution of the Common Palestinian-Israeli Waters. Pp. 451-467 in Proceedings of the First Israeli-Palestinian International Academic Conference on Water, Water and Peace in the Middle East, J. Isaac and H. Shuval, eds., Zurich, Switzerland, December 10-13, 1992. New York, N.Y.: Elsevier.

Environmental Protection Agency. 1983. Process design manual: Land application of municipal sludge. EPA 625/1-83-016. Cincinnati, OH: Center for Environmental Research Information.

Environmental Protection Agency. 1984. Process design manual: Land treatment of municipal wastewater. EPA 625/1-81-013. Cincinnati, OH: Center for Environmental Research Information.

Environmental Protection Agency. 1984. Environmental regulations and technology—Use and disposal of municipal wastewater sludge. EPA 625/1-81-013a. Cincinnati, OH: Environmental Research Information.

Ergin, M., D. Altinbilek, and M. R. Zou'bi, eds. 1994. Water in the Islamic World: An Imminent Crisis. In Proceedings of the Conference on Water in the Islamic World: An Imminent Crisis, Khartoum, Sudan, 5-9 December. Amman, Jordan: The Islamic Academy of Sciences.

Farid, A. M., and H. Sirriyeh. 1985. Israel and Arab Water: An International Symposium, February 25-26, 1984, Amman, Jordan.

Felgin, A., I. Ravina, and J. Shalhevet. 1991. Irrigation with Treated Sewage Effluent—Management for Environmental Protection. Advance Series in Agricultural Sciences 17. Berlin: Springer Verlag. P. 224.

Feitelson, E., and M. Haddad. 1995. Joint Management of Shared Aquifers: Final Report. Cooperative Research Project: Palestine Consultancy Group, East Jerusalem, and Truman Research Institute for Advancement of Peace, Hebrew University of Jerusalem. 32 pp.

Fischer, S., D. Rodrik, and E. Tuma, eds. 1993. The Economics of Middle East Peace: Views from the Region. Cambridge, Mass.: MIT Press.

Fisher, F. M. 1995. The Economics of Water Dispute Resolution, Project Evaluation and Management: An Application to the Middle East. Water Resources Development 11:377-390.

Ghazi, F. 1990. Arabs vs Jews in Galilee: Competition for regional resources. GeoJournal 21(4):325-336.

Gleik, P. H. 1994. Reducing the risks of conflict over fresh water resources in the Middle East. Pp. 41-54 in Proceedings of the First Israeli-Palestinian International Academic Conference on Water, Water and Peace in the Middle East, J. Isaac and H. Shuval, eds., Zurich, Switzerland, December 10-13, 1992. New York, N.Y.: Elsevier.

Gleik, P. H. 1994. Water, war and peace in the Middle East. Environment 36(3):6-42.

Gleik, P. H. 1998. The World's Water, 1998-1999. Covelo: Island Press.

Gray, J. F. 1993. The Importance of Agriculture to Wastewater Reclamation and Reuse Systems, History of the Reclamation and Reuse System of Lubbock, Texas. City of Lubbock, Texas.

Gruen, G. E. 1994. Contribution of Water Imports to Israeli-Palestinian-Jordanian Peace. Pp. 273-288 in Proceedings of the First Israeli-Palestinian International Academic Conference on Water, Water and Peace in the Middle East, J. Isaac and H. Shuval, eds., Zurich, Switzerland, December 10-13, 1992. New York, N.Y.: Elsevier.

Gur, S. 1992. A View From Israel. Pp. 1-13 in Water and the Peace Process: Two Perspectives—Policy Focus. Research Memorandum #20, September 1992. Washington, D.C.: Washington Institute for Near East Policy.

Haddadin, M. 1992. A View From Jordan. Pp. 1, 14-19 in Water and the Peace Process: Two Perspectives—Policy Focus. Research Memorandum #20, September 1992. Washington, D.C.: Washington Institute for Near East Policy.

Haimes, Y. Y. 1992. Sustainable Development: A Holistic Approach to Natural Resources Management. Water International 17:187-192.

Hammond, A., A. Adriaanse, E Rodenburg, D. Bryant, and R. Woodward. 1995. Environmental Indicators: A Systematic Approach to Measuring and Reporting on Environmental Policy Performance in the Context of Sustainable Development. Washington, D.C.: World Resources Institute.

Harry S. Truman Research Institute for the Advancement of Peace. 1995. Joint Management of Shared Aquifers: The Second Workshop, November 27 December 1, 1994. M. Haddad and E. Feitelson, eds. Jerusalem, Israel: The Hebrew University of Jerusalem.

Hayes, J. B., and A. E. Barrekette. 1948. T.V.A. on the Jordan: Proposals for Irrigation and Hydro-Electric Development in Palestine. A report prepared under the auspices of the Commission on Palestine Surveys. Washington, D.C.: Public Affairs Press.

Hillel, D. 1994. Rivers of Eden. New York, N.Y.: Oxford University.

Hinman, C., and J. Hinman. 1992. The Plight and Promise of Arid Land Agriculture. New York: Columbia University.

Hoffman, D. 1994. Potential Applications for Desalination in the Area. Pp. 315-327 in Proceedings of the First Israeli-Palestinian International Academic Conference on Water, Water and Peace in the Middle East, J. Isaac and H. Shuval, eds., Zurich, Switzerland, December 10-13, 1992. New York, N.Y.: Elsevier.

Hydrological Service of Israel. 1995. Exploitation and State of Groundwater Resources in Israel until Fall 1994. P. 199 (in Hebrew).

International Law Association. 1986. Helsinski Tules on the Uses of the Waters of International Rivers. Adopted by the International Law Association at the 52nd Conference held in Helsinski, 30 August 1986. London, U.K.: International Law Association.

International Water Engineering Center. Enhancement of Middle East Water Supply: A Literature Survey of Technologies and Applications, E. M. Lentz and E. Andras, eds., 2nd ed. Ontario, Canada: University of Ottawa.

International Water Engineering Center. Enhancement of Middle East Water Supply: A Literature Survey of Technologies and Applications. 1993. Report and Appendix A. Ontario, Canada: University of Ottawa.

International Water Resources Association. 1993. Water in the Middle East. Special edition, Water International 18:1.

Irrigation Support Project for Asia and the Near East. 1992. Mid-East Regional Water Schemes. Draft. U.S. Department of State and U.S. Agency for International Development.

Isaac, J., and H. Shuval, eds. 1994. Water and Peace in the Middle East. Proceedings of the First Israeli-Palestinian International Academic Conference on Water, Zurich, Switzerland, December 10-13, 1992. New York, N.Y.: Elsevier.

Jordan Information Bureau. 1991 & 1992. Water the future challenge: Face-to-face with the frightening reality of water shortages. Jordan Issues and Perspectives.

Kahhaleh, S. 1981. The Water Problem in Israel and its Repercussions on the Arab-Israeli conflict. Beirut: Institute for Palestine Studies.

Kally, E. 1993. Water and Peace: Water resources and the Arab-Israeli peace process. Westport, Conn: Praegar.

Kanarek, A., A. Aharoni, M. Michail, I. Kogan, and D. Sherer. 1994. Dan Region Reclamation Project, Groundwater Recharge with Municipal Effluent. Tel Aviv, Israel: Mekorot Water Company Ltd. P. 150 (in Hebrew).

Kay, F. A., and B. Mitchell. 1998. Performance of Israel's Water System under a New Minister Plan: Post-audit and Implications for the Future. Water Resources Development 14:107.

Kliot, N. 1993. Water Resources and Conflict in the Middle East. London and New York: Routledge.

Loehr, R. C., W. J. Jewell, J. D. Novak, W. W. Clarkson, and G. S. Friedman. 1979. Land Application of Wastes. Vol. I and II. New York, N.Y: Van Nostrand Reinhold.

Loehr, R. C., and M. R. Overcash. Land treatment of wastes: Concepts and general design. J. Environ. Engr. Div. ASCE III:141-160.

Lonegran, S. C., and D. B. Brooks. 1994. Watershed: The role of fresh water in the Israeli-Palestinian conflict. Ontario, Canada: International Development Research Center.

Lowi, M. R. 1993. Bridging the divide: Transboundary resource disputes and the case of West Bank water. International Security 18(1):113-138.

Lowi, M. R. 1993. Water and Power: The politics of a scarce resource in the Jordan River basin. Cambridge: Cambridge University.

Mendelsohn, E. 1989. A Compassionate Peace: A Future for Israel, Palestine, and the Middle East. New York, N.Y.: Farrar, Straus, & Giroux.

Mediterranean Commission for Sustainable Development. 1997. Main Facts and Figures on Water Demands in the Mediterranean Region in Workshop on Water Demands Management, 12-13 September 1997. Mediterranean Action Plan.

Mekorot Water Co. Ltd. 1991. Dan Region Sewage Treatment and Reclamation Project. P. 20.

MEWIN. 1996. Directory: Individuals and Organizations Specializing in Middle East Water Resources. Philadelphia, Penn.: Osage Press.

Moldan, B., S. Billharz, and R. Matravers, eds. 1997. Sustainability Indicators: A Report on the Project on Indicators of Sustainable Development. Scope 58. Chichester: J. Wiley & Sons.

Morris, M. E. 1993. Dividing the waters: Reaching equitable water solutions in the Middle East. Santa Monica, Calif.: Rand Library Collection.

Morrison, J. I., S. L. Postel, and P. H. Gleich. 1996. The Sustainable Use of Water in the Lower Colorado River Basin: A joint report of Pacific Institute and the Global Water Project. Nairobi: UNEP.

Murakami, M., and A. T. Wolf. 1995. Techno-political water and energy development alternatives in the Dead Sea and Aqaba regions. Water Res. Dev. 11(2):163-183.

Murakami. M. 1995. Managing Water for Peace in the Middle East: Alternative Strategies. The New York and Tokyo, Japan: U.N. University Press.

Musallam, R. 1990. Water: Source of conflict in the Middle East in the 1990s. London: Gulf Centre for Strategic Studies.

Mustafa, I. 1994. The Arab-Israeli conflict Over Water Resources. Pp. 123-133 in Proceedings of the First Israeli-Palestinian International Academic Conference on Water, Water and Peace in the Middle East, J. Isaac and H. Shuval, eds., Zurich, Switzerland, December 10-13, 1992. New York, N.Y.: Elsevier.

Naff, T. 1997. A Selected Bibliography of Sources on the International Law of Fresh Water Resources. Middle East Water Information Network (MEWIN), University of Pennsylvania, Philadelphia.

Naff, T. 1997. Information systems, water, and conflict: Exploring the linkages in the Middle East. Water International 32.

Naff, T. 1987. The potential and limits of technology, Part I. Associates for Middle East Research, Water Project.

Naff, T., and R. C. Matson, eds. 1984. Water in the Middle East: Conflict or Cooperation? Published in cooperation with Middle East Research Institute, University of Pennsylvania. Boulder, Colo.: Westview.

Nasr, R., and R. A. Al-Weshah. 1993. Optimizing the Irrigation Water Use for Vegetable Production in the Jordan Valley: A Case Study. Pp. 173-180 in Proceedings of the International Symposium on Water Resources in the Middle East. IWRI publications. University of Illinois at Urbana-Champaign, October 25-28.

National Research Council. 1977. Drinking Water and Health. Washington, D.C.: National Academy Press.

National Research Council. 1980. Drinking Water and Health, Volume 2. Washington, D.C.: National Academy Press.

National Research Council. 1980. Drinking Water and Health, Volume 3. Washington, D.C.: National Academy Press.

National Research Council. 1982. Drinking Water and Health, Volume 4. Washington, D.C.: National Academy Press.

National Research Council. 1982. Quality Criteria for Water Reuse. Washington, D.C.: National Academy Press.

National Research Council. 1983. Drinking Water and Health, Volume 5. Washington, D.C.: National Academy Press.

National Research Council. 1986. Drinking Water and Health, Volume 6. Washington, D.C.: National Academy Press.

National Research Council. 1986. Ecological Knowledge and Environmental Problem Solving: Concepts and Case Studies. Washington, D.C.: National Academy Press.

National Research Council. 1987. Drinking Water and Health, Disinfectants and Disinfectant By-Products, Volume 7. Washington, D.C.: National Academy Press.

National Research Council. 1987. Drinking Water and Health, Pharmacokinetics in Risk Assessment, Volume 8. Washington, D.C.: National Academy Press.

National Research Council. 1987. The Mono Basic Ecosystem: Effects of Changing Lake Level. Washington, D.C.: National Academy Press.

National Research Council. 1988. Hazardous Waste Site Management: Water Quality Issues. Washington, D.C.: National Academy Press.

National Research Council. 1989. Irrigation-Induced Water Quality Problems. Washington, D.C.: National Academy Press.

National Research Council. 1990. Ground Water and Soil Contamination Remediation. Washington, D.C.: National Academy Press.

National Research Council. 1992. Restoration of Aquatic Ecosystems. Washington, D.C.: National Academy Press.

National Research Council. 1992. Water Transfers in the West. Washington, D.C.: National Academy Press.

National Research Council. 1993. In-Situ Bioremediation. Washington, D.C.: National Academy Press.

National Research Council. 1993. Ground Water Vulnerability Assessment. Washington, D.C.: National Academy Press.

National Research Council. 1994. Ground Water Recharge Using Waters of Impaired Quality. Washington, D.C.: National Academy Press.

National Research Council. 1994. Alternatives for Ground Water Cleanup. Washington, D.C.: National Academy Press.

National Research Council. 1995. Mexico City's Water Supply: Improving the Outlook for Sustainability. Washington, D.C.: National Academy Press.

National Research Council. 1995. Wetlands: Characteristics and Boundaries. Washington, D.C.: National Academy Press.

Nativ, R. 1988. Problems of an Over-Developed Water System: The Israeli Case. Water Quality Bulletin 13:13-26.

Okun, D. 1994. The Role of Reclamation and Reuse in Addressing Community Water Needs in Israel and the West Bank. Pp. 329-338 in Proceedings of the First Israeli-Palestinian International Academic Conference on Water, Water and Peace in the Middle East, J. Isaac and H. Shuval, eds., Zurich, Switzerland, December 10-13, 1992. New York: Elsevier.

Overcash, M. D., and D. Pal. 1979. Design of Land Treatment Systems for Industrial; Wastes—Theory and Practice. Stoneham, Mass.: Ann Arbor Science/Butterworths.

Page, A. L., T. L. Gleason, J. E. Smith, J. K. Iskander, and L. E. Sommers. 1983. Utilization of Municipal Wastewater and Sludge on Land. Riverside: University of California.

Pescod, M. B. 1992. Wastewater treatment and use in Agriculture. FAO Irrigation and Drainage Paper 47. Rome: FAO. P. 125.

Pettygrove, G. S., and T. Asano, eds. 1985. Irrigation with Reclaimed Municipal Wastewater—A Guidance Manual. Chelsea. MI: Lewis Publishers.

Rabr, A., F. Daibes, and A. Aliewi. 1994. Availability and Reliability of Secondary Source Hydrogeological Data for the West Bank with Additional Reference Material for Gaza Strip, Jerusalem. Water Resources Management: West Bank and Gaza Strip.

Ragheb, M., K. Toukan, and R. A. Al-Weshah. 1993. Desalination Using Advanced-Design Nuclear Power Plants. Pp. 173-180 in Proceedings of the International Symposium on Water Resources in the Middle East. IWRA publications. University of Illinois at Urbana-Champaign, October 25-28.

Sbeih, M. Y. 1994. Reuse of Waste Water for Irrigation in the West Bank: Some Aspects. Pp. 339-350 in Proceedings of the First Israeli-Palestinian International Academic Conference on Water, Water and Peace in the Middle East, J. Isaac and H. Shuval, eds., Zurich, Switzerland, December 10-13, 1992. New York, N.Y.: Elsevier.

Scharlat, R. R. 1994. Arab-Israeli water resources: Tapping cooperation, diluting conflict. Master's thesis, Sever Institute of Technology, Washington University, St. Louis, MO.

Sheaffer, J. R. The Modular Reclamation and Reuse Technology: An Approach to Sustainability. Sheaffer International, Ltd., Integrated Water Resources Management, Naperville, Illinois.

Sheaffer, J. R. Wastewater Reclamation and Reuse Systems: A Forgotten Planning Tool. Presented at the Uniersity of Illinois, Champaign.

Sheaffer, J. R. 1984. Going back to Nature's Way: Circular vs. Linear Water Systems. Environment October 15:42-44.

Sheaffer, J. R. 1979. Land Application of Waste - Important Alternative. Ground Water 17(1)January-February:62-68.

Shechter, M., ed. 1994. Sharing Water Resources in the Middle East, Economic Perspective. Resource and Energy Economics 16(4)265-389.

Schiffler, M. et al. 1994. Water Demand Management in an Arid Country: The Case of Jordan With Special References to Industry. Berlin: German Development Institute.

Schmida, L. C. 1983. Keys to control: Israel's pursuit of Arab water resources. Washington, D.C.: American Educational Trust.

Shahin, M. 1989. Review and Assessment of Water Resources in the Arab Region. Water International 14(4):206-219.

Shalhevet, J., A. Mantell, H. Bielorai, and D. Shimshi. 1981. Irrigation of Field and Orchard Crops under Semi-Arid Conditions. International Irrigation Information Center Publ. No. 1. Bet Dagan, Israel. Pp. 132.

Soffer, A. 1994. The Relevance of Johnston Plan to the Reality of 1993 and Beyond. Pp. 107-121 in J. Isaac and H. Shuval, eds. Water and Peace in the Middle East. Proceedings of the First Israeli-Palestinian International Academic Conference on Water, Zurich, Switzerland, December 10-13, 1992. New York: Elsevier.

Starr, J. R. 1991. Water wars. Foreign Policy 82(Spring):17-36.

Starr, J. R., and D. C. Stoll, eds. 1988. The Politics of Scarcity: Water in the Middle East. Boulder, CO: Westview.

Stout, G. E., and R. A. Al-Weshah, eds. 1993. Proceedings of the International Symposium on Water Resources in the Middle East. IWRA publications. University of Illinois at Urbana-Champaign, October 25-28, 285 pp.

Tahal Consulting Eng. Ltd. 1993. Israel Water Study for the World Bank. A draft report.

The Wrap Task Force. 1994. A Rapid Interdisciplinary Sector Review and Issues Papers. Task Force of the Water Resources Action Program.

United Nations Commission on Sustainable Development. 1997. Comprehensive Assessment of the Freshwater Resources of the World. Report of the Secretary General. New York. 52 pp.

United Nations Environment Program and Twenty Cooperating Research Centres. 1997. Global Environment Outlook. New York: Oxford University, Paris. 264 pp.

United Nations Food and Agriculture Organization. 1997. Proceedings of the Expert Consultation on National Water Policy Region in the Near East, Beirut, Lebanon, 9-10 December 1996. Cairo: FAO Regional Office.

Van Tuijl, W. 1993. Improving water use in agriculture: Experiences in the Middle East and North Africa. World Bank Technical Paper Number 201, July.

Vengosh, A., and E. Rosenthal. 1993. Saline ground water and its influence on water quality in Israel. Hydrological Service of Israel, Jerusalem. Rep. 1993/6. P. 30 (in Hebrew).

Vesilind, P. J. 1993. The Middle East's WATER critical resource. National Geographic 183(5):38-70.

Wachtel, B. 1994. The Peace Canal Project: A Multiple Conflict Resolution Perspective for the Middle East. Pp. 363-374 in Proceedings of the First Israeli-Palestinian International Academic Conference on Water, Water and Peace in the Middle East, J. Isaac and H. Shuval, eds., Zurich, Switzerland, December 10-13, 1992. New York, N.Y.: Elsevier.

Waterbury, J. 1992. Three Rivers in Search of a Regime: The Jordan, the Euphrates, and the Nile. Proceedings. Paper presented at the 17th Annual Symposium of the Center for Contemporary Arab Studies. Washington, D.C.: Georgetown University.

Water Science and Technology Board. 1992. Irrigation: A Blessing or a Curse? Transcript of Abel Wolman Lecture by Jan van Schilfgaarde. Washington, D.C.: National Research Council.

Water Science and Technology Board. 1993. Transnational Water Resources Management: Learning from the Mexico Example. Transcript from 1993 Wolman Lecture by Helen Ingram. Washington, D.C.: National Research Council.

Wolf, A. T. 1992. The impact of scarce water resources on the Arab-Israeli conflict: An interdisciplinary study of water conflict analysis and proposals for conflict resolution. Ph.D. Dissertation, Dept. of Land Resources, University of Wisconsin-Madison.

Wolf, A. T. 1992. Impact of Scarce Water Resources on the Arab-Israeli Conflict. Natural Resources J. 32(4).

Wolf, A. T. 1992. Hydropolitics Along the Jordan River. New York, N.Y.: United Nations University Press.

Wolf, A. T. 1993. Water for peace in the Jordan river watershed. Natural Resources J. 33(3):797-839.

Wolf, A. T. 1995. Hydropolitics Along the Jordan River—Scarce Water and Its Impacts on the Arab-Israeli Conflict. New York and Tokyo, Japan: U.N. University Press.

Wolf, A. T. 1995. International Water Dispute Resolution: The Middle East multilateral working group on water resources. Water International 20:141-150.

Wolf, A. T., and J. Ross. 1992. The impact of scarce water resources on the Arab-Israeli conflict. Natural Resources J. 32(Fall):919-958.

Wolf, A. T., and M. Murkami. 1995. Techno-political decision making for water resources development: The Jordan river watershed. Water Res. Dev. 11(2):147-162.

Wolf, A. T., and S. Lonegran. 1995. Pp. 179-187 in Resolving Conflicts Over Water Disputes in the Jordan River Basin, A. Dinar and E.T. Loehman, eds. Westport, Conn.: Praeger Publishers.

World Bank, Water Resource Management Unit. 1993. A Strategy for Managing Water in the Middle East and North Africa. ECA/MENA Technical Department, The World Bank, September 20.

Zahlan, A., ed. 1985. The Agricultural Sector of Jordan: Policy & systems studies. London, U.K.: Ithaca Press.

Zarour, H., and J. Isaac. 1993. Nature's Apportionment and the Open Market: A promising solution to the Arab-Israeli Water Conflict. Water International 18:40-53.

APPENDIX
F

List of Meeting Places and Dates

First Committee Meeting
February 14-16, 1996
Washington, D.C.

Second Committee Meeting
June 17-19, 1996
Amman, Jordan

Third Committee Meeting
September 18-20, 1996
Haifa, Israel

Fourth Committee Meeting
April 2-4, 1997
Washington, D.C.

Dissemination Meeting
March 2, 1999
Ramallah, West Bank

Committee on Sustainable Water Supplies for the Middle East Biographical Sketches

GILBERT F. WHITE received his S.B., S.M., and Ph.D. in geography from the University of Chicago in 1932, 1934, and 1942, respectively. His research covers natural resources management, environmental policy, and natural hazards. He has served as vice chair of the U.S. President's Water Resources Policy Commission, president of the Scientific Committee on Problems of the Environment (SCOPE), and chair of various water studies, including a review for the Lower Mekong Commission, the United Nations (UN) Development Programme review of African water projects, and the UN Environment Programme diagnostic study of the Aral Sea Basin. Presently, he is distinguished professor emeritus of geography at the University of Colorado, Boulder, and a member of the National Academy of Sciences.

YOUSIF ATALLA ABU SAFIEH received his B.S. in biology-chemistry in 1972 and his M.Sc. in parasitology in 1977; both from the American University of Beirut. He received his Ph.D. in environmental sciences from the University of Texas, Houston, in 1986. He is currently Minister of State for the Environment. He is also acting vice president for academic affairs at Al-Azhar University, Gaza, and a core team member and water quality expert for the Gaza Environmental Profile. Dr. Abu-Safieh is a member of the International Society of Environmental Epidemiology.

AYMAN A. AL-HASSAN holds an M.S. in chemical engineering. Since 1983 he has held various positions at the Royal Scientific Society as a researcher, head of the air pollution unit, and of the air pollution and hazardous chemicals division. Currently he is director of the Environmental Research Center. As director, he manages, plans, and directs research work and studies related to enhancement of water quality and prevention of water resources degradation, assessment of air quality, and solid and hazardous waste management.

RADWAN A. AL-WESHAH has a Ph.D. in civil engineering-water resources and hydrology from the University of Illinois at Urbana-Champaign. He is an assistant professor of hydrology and water resources at the Department of Civil Engineering and a researcher at the Water and Environment Research and Study Center, University of Jordan. He is involved in many projects related to water resources, hydrologic studies, and hydrologic/ hydraulic design in Jordan and abroad. Dr. Al-Weshah has international experience in water resources management and planning, surface hydrology, wetland hydrology, and water resource issues in the Middle East. He was the co-chair and main coordinator for the International Symposium on Water Resources in the Middle East, held at the University of Illinois in 1993.

KAREN ASSAF received her B.S. in science-geology and her M.S. in earth science-geology from Iowa State University in 1967 and 1968, respectively. She received her Ph.D. in environmental science-hydrology from the University of Texas in 1976. Her specialties and interests include environmental science-/hydrology, public health, and statistics. Dr. Assaf currently is director of the Department of Water Planning at the Ministry of Planning and International Cooperation, working in the Palestinian Water Authority, Palestinian Interim Self-Government Authority. Previously she was research scientist and director of Water Division and Publications at the Arab Scientific Institute for Research and Transfer of Technology, where she conducted studies in water, health, food, and nutrition. From 1979 to 1987 she was project coordinator for the Water, Sanitation and Health Projects for Save the Children where she directed the basic needs sector (water, sanitation, and health) for projects funded by the U.S. Agency for International Development in the West Bank and Gaza Strip. Dr. Assaf is a board member of the International Water Resources Association.

YORAM AVNIMELECH received his M.Sc. in soil microbiology and his Ph.D. in soil physical chemistry from the Weizmann Institute and the Hebrew University of Jerusalem. He is currently S. Gurney chair and professor of agricultural engineering at the Technion, Israel Institute of Technology. His academic background led to strong multidisciplinary interests, covering topics from microbiology to physical chemistry, and from agronomy to health studies, with a specialty in soil sciences. From 1989 to 1994, he was chief scientist of the Israeli Ministry of the Environment. He published more than 100 scientific papers in the environmental sciences.

EDITH BROWN WEISS received her A.B. from Stanford University, LL.B. from Harvard Law School, and Ph.D. from the University of California at Berkeley. She is Francis Cabell Brown Professor of International Law at Georgetown University Law Center, where she has taught international law, international environmental law, water law, and environmental law. From 1990 to 1992, she was on leave to the U.S. Environmental Protection Agency, where she served as associate general counsel for International Activities and established a new division to address international legal issues. Dr. Brown Weiss was president of the American Society of International Law from 1994 to 1996, and chaired the Social Science Research Council's Committee on Research in Global Environmental Change from 1989 to 1994. She is a special legal advisor to the North American Commission on Environmental Cooperation and a member of the American Law Institute and the Council on Foreign Relations. Her book, *In Fairness to Future Generations*, received the Certificate of Merit from the American Society of International Law in 1990. She received the Elizabeth Haub Prize awarded by the Free University of Brussels and the International Council on Environmental Law for exceptional achievement in International Environmental Law. Dr. Brown Weiss was a member of the U.S. National Research Council's (NRC) Water Science and Technology Board from 1985 to 1988 and has served on several NRC committees. She was a member of the NRC's Commission on Geosciences, Environment, and Resources from 1992 to 1995.

CHARLES D. D. HOWARD is a civil engineering graduate of the University of Alberta (B.Sc. in 1960 and M.Sc. in 1962), and the Massachusetts Institute of Technology (M.S., 1966). Since 1975, he has been president of the water resources engineering firm Charles Howard and Associates, Ltd., specializing in applications of systems analysis. He is a consultant to water sewerage and hydroelectric power utilities, provincial, state, and federal governments in Canada, the United States, and Mexico, and to the World Bank. From 1986 to 1996 he was a member of an NRC committee that addressed irrigation-induced water quality problems in the western United States. He is a member of the NRC's Water Science and Technology Board.

IRWIN H. KANTROWITZ holds degrees in geology from Brooklyn College (B.S., in 1958) and the Ohio State University (M.S., in 1959). He retired from the U.S. Geological Survey (USGS) in 1995, after serving as a geologist and hydrologist for 36 years. He conducted and directed hydrologic resource appraisal and research programs in New York, Maryland, and Florida and was a member of the Water Resources Science Advisory Committee. He also served as a member of the USGS Ground-Water Quality Delegation to the People's Republic of China and as a member of the Middle East Water Data Banks Delegation as part of the Multilateral

Working Group on Water Resources. He is a recipient of the U.S. Department of the Interior's Meritorious and Distinguished Service awards.

RAYMOND LOEHR received his B.S. and M.S. from Case Institute of Technology in 1953 and 1956, respectively. He received his Ph.D. in sanitary engineering from the University of Wisconsin in 1961. His area of expertise is environmental engineering. His research includes environmental health engineering, water and wastewater treatment, hazardous waste treatment, industrial waste management, and land treatment of wastes. Currently he is H. M. Alharthy Centennial Chair and Professor of Civil Engineering at the University of Texas at Austin. Previously, he was professor of agriculture engineering and civil engineering at Cornell University. He has been a consultant to numerous industries, agencies, and engineering firms on matters dealing with industrial and hazardous waste management, particularly in the petroleum, petrochemical, food processing, and pulp and paper industries. He is a member of the U.S. National Academy of Engineering.

AYMAN I. RABI received his B.S. in hydrogeological engineering from Dokuz Eylul University at Izmir, Turkey, in 1987. He received his M.S. in water resource systems engineering from the University of Newcastle Upon Tyne in 1993. He is a water resource systems engineer. He has more than six years of experience in rural development issues and strategies, especially those related to water resources development. Mr. Rabi is the founder of the Palestinian Hydrology Group (PHG). He has supervised the execution of most of the water projects implemented by the group over the past seven years. He also has experience in institutional management practices and currently is acting as the executive manager of the PHG. Mr. Rabi is currently supervising a research project to establish water information and water resource monitoring systems in the West Bank. This project is considered the first real project that will gather and process various types of data related to water resources planning and management in Palestine.

URIEL N. SAFRIEL received his M.Sc. in zoology from The Hebrew University of Jerusalem in 1964 and his D.Phil. in ecology from Oxford University in 1967. Since 1969 he has been on the faculty of The Hebrew University of Jerusalem. Since 1983 he has been a visiting scientist (part time) at Ben Gurion University, where he has been head of the Mitrani Center for Desert Ecology at the Blaustein Institute for Desert Research. Since 1974 he has been curator of birds at the Museum of Zoology at The Hebrew University of Jerusalem. He was chief scientist and director of the Science and Management Division at the Israel Nature Reserves Authority from 1989 to 1991. Since 1992 he has been a member of the National Committee for Climate Change, an appointment by the Ministry of the Environment, and since 1993 he has represented Israel in the Intergovernmental Negotiating Committee on the Convention to Combat Desertification. Since 1996 he has been the director of the Blaustein Institute for Desert Research.

ELIAS SALAMEH obtained his D.Sc. in hydrogeology and hydrochemistry from the Technical University of Munich, Germany, in 1974. Dr. Salameh is a professor of hydrogeology and hydrochemistry at the University of Jordan. In 1982-1983, he headed the Department of Geology at the University of Jordan, and in 1983 founded the Water Research and Study Center, serving as its director from 1983 to 1992. In 1990, he chaired the Water and Agriculture Sector of the Higher Council of Science and Technology in Jordan. From 1975 until 1992, he published more than 100 articles in hydrogeology, hydrochemistry, and water resources of the Middle East, primarily in international journals.

JOSEPH SHALHEVET received his B.S. in agriculture from the University of California, Berkeley, in 1954. He received his M.S. and Ph.D. in soil science from Cornell University in 1955 and 1958, respectively. Currently he is director of the Jordan Valley Research and Development Organization. In 1979, he was awarded the first Y. Ratner Award given by the Agricultural Research Organization in Israel for outstanding achievement in soils and water research. He was the recipient of the 1989 American Society of Agronomy Fellow Award and the 1989 Soil Science Society of America Fellow Award. He served as chief scientist and director of the Agricultural Research Organization of the Ministry of the Agriculture from 1986 to 1990. From 1990 to 1992 he was the director of the Liaison Office of the Israel Academy of Sciences and Humanities in Beijing, China, and following the establishment of diplomatic relations between China and Israel, served as counsellor for Science and Technology at the Embassy of Israel, Beijing (1992). He was chairman of the board of directors of the U.S. Israel Binational Agricultural Research and Development fund (1987-1988) and chairman of the board of directors and scientific consultant to the International Irrigation Information Center from 1976 to 1986.

HENRY VAUX received a Ph.D in economics from the University of Michigan in 1973. He is professor of resource economics at the University of California, Riverside. He currently serves as associate vice president of agricultural and natural resource programs for the University of California system. He previously served as director of the University of California Water Resource Center. His principal research interests are the economics of water use and water quality. Prior to joining the University of California, he worked at the Office of Management and Budget and served on the staff of the National Water Commission. Dr. Vaux has been a member of the NRC's Water Science and Technology Board since 1994 and is its current chair.

NATIONAL RESEARCH COUNCIL STAFF

SHEILA D. DAVID serves as a consultant for the Water Science and Technology Board's Committee on Sustainable Water Supplies for the Middle East. Currently she is a fellow and project manager at the H. John Heinz III Center for Science, Economics and the Environment. Until August 1997, she was a senior program officer at the Water Science and Technology Board. On the staff of the National Research Council NRC since 1976, she served as staff director for approximately 30 NRC study projects, including studies on coastal erosion, ground-water protection, water quality and water reuse, wetlands, and natural resources protection and river management in the Grand Canyon.

DAVID POLICANSKY has a B.A. in biology from Stanford University and an M.S. and Ph.D., biology, from the University of Oregon. He is associate director of the Board on Environmental Studies and Toxicology at the National Research Council. Formerly, he taught and conducted research at the University of Chicago, the University of Massachusetts at Boston, and the Grey Herbarium of Harvard University. He was a visiting scientist at the National Marine Fisheries Service Northeast Fisheries Center. He is a member of the Ecological Society of America and the American Fisheries Society, and chairs the advisory councils to the University of Alaska's School of Fisheries and Ocean Sciences and the University of British Columbia's Fisheries Centre. He was a member of the editorial board of *BioScience*. His interests include genetics, evolution, and ecology, particularly the effects of fishing on fish populations, ecological risk assessment, and natural resource management. He has directed approximately 20 projects at the NRC on natural resources and ecological risk assessment.

JEANNE AQUILINO is the NRC administrative specialist for the Water Science and Technology Board and the Committee on Sustainable Water Supplies for the Middle East. She has been on the NRC staff since 1979, helping to manage a wide variety of studies, including Restoration of Aquatic Ecosystems, Managing Coastal Erosion, and Ground Water Models: Scientific and Regulatory Applications.